# MATH LAB

**TASK CARD SERIES**

Conceived and written by

**RON MARSON**

Illustrated by

**PEG MARSON**

TOPS LEARNING SYSTEMS

342 S Plumas Street
Willows, CA 95988

www.topscience.org

# WHAT CAN YOU COPY?

Dear Educator,

   Please honor our copyright restrictions. We offer liberal options and guidelines below with the intention of balancing your needs with ours. When you buy these labs and use them for your own teaching, you sustain our work. If you "loan" or circulate copies to others without compensating TOPS, you squeeze us financially, and make it harder for our small non-profit to survive. Our well-being rests in your hands. Please help us keep our low-cost, creative lessons available to students everywhere. Thank you!

# PURCHASE, ROYALTY and LICENSE OPTIONS

## TEACHERS, HOMESCHOOLERS, LIBRARIES:

   We do all we can to keep our prices low. Like any business, we have ongoing expenses to meet. We trust our users to observe the terms of our copyright restrictions. While we prefer that all users purchase their own TOPS labs, we accept that real-life situations sometimes call for flexibility.

   *Reselling, trading, or loaning our materials is prohibited* unless one or both parties contribute an Honor System Royalty as fair compensation for value received. We suggest the following amounts – let your conscience be your guide.

   *HONOR SYSTEM ROYALTIES:* If making copies from a library, or sharing copies with colleagues, please calculate their value at 50 cents per lesson, or 25 cents for homeschoolers. This contribution may be made at our website or by mail (addresses at the bottom of this page). Any additional tax-deductible contributions to make our ongoing work possible will be accepted gratefully and used well.

   Please follow through promptly on your good intentions. Stay legal, and do the right thing.

## SCHOOLS, DISTRICTS, and HOMESCHOOL CO-OPS:

*PURCHASE Option:* Order a book in quantities equal to the number of target classrooms or homes, and receive quantity discounts. If you order 5 books or downloads, for example, then you have unrestricted use of this curriculum for any 5 classrooms or families per year for the life of your institution or co-op.

   **2-9 copies of any title:** 90% of current catalog price + shipping.

   **10+ copies of any title:** 80% of current catalog price + shipping.

*ROYALTY/LICENSE Option:* Purchase just one book or download *plus* photocopy or printing rights for a designated number of classrooms or families. If you pay for 5 additional Licenses, for example, then you have purchased reproduction rights for an entire book or download edition for any **6** classrooms or families per year for the life of your institution or co-op.

   **1-9 Licenses:** 70% of current catalog price per designated classroom or home.

   **10+ Licenses:** 60% of current catalog price per designated classroom or home.

## WORKSHOPS and TEACHER TRAINING PROGRAMS:

   We are grateful to all of you who spread the word about TOPS. Please limit copies to only those lessons you will be using, and collect all copyrighted materials afterward. No take-home copies, please. Copies of copies are strictly prohibited.

   For licensing, honor system royalty payments, contact: **www.TOPScience.org**; or **TOPS Learning Systems 342 S Plumas St, Willows CA 95988**; or inquire at **customerservice@topscience.org**

## ISBN 978-0-941008-77-8

# CONTENTS

# A TOPS Model for Effective Science Teaching...

*If science were only a set of explanations* and a collection of facts, you could teach it with blackboard and chalk. You could assign students to read chapters and answer the questions that followed. Good students would take notes, read the text, turn in assignments, then give you all this information back again on a final exam. Science is traditionally taught in this manner. Everybody learns the same body of information at the same time. Class togetherness is preserved.

**But science is more than this.**

Science is also process — a dynamic interaction of rational inquiry and creative play. Scientists probe, poke, handle, observe, question, think up theories, test ideas, jump to conclusions, make mistakes, revise, synthesize, communicate, disagree and discover. Students can understand science as process only if they are free to think and act like scientists, in a classroom that recognizes and honors individual differences.

Science is *both* a traditional body of knowledge *and* an individualized process of creative inquiry. Science as process cannot ignore tradition. We stand on the shoulders of those who have gone before. If each generation reinvents the wheel, there is no time to discover the stars. Nor can traditional science continue to evolve and redefine itself without process. Science without this cutting edge of discovery is a static, dead thing.

Here is a teaching model that combines the best of both elements into one integrated whole. It is only a model. Like any scientific theory, it must give way over time to new and better ideas. We challenge you to incorporate this TOPS model into your own teaching practice. Change it and make it better so it works for you.

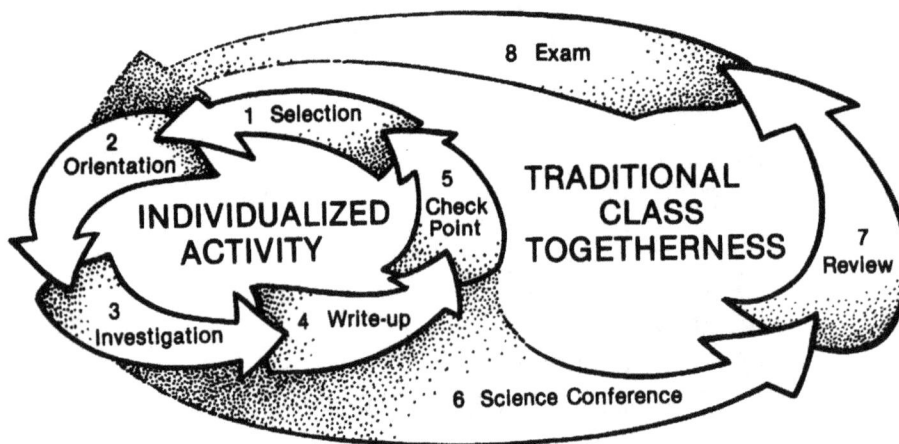

## 1. SELECTION

Doing TOPS is as easy as selecting the first task card and doing what it says, then the second, then the third, and so on. Working at their own pace, students fall into a natural routine that creates stability and order. They still have questions and problems, to be sure, but students know where they are and where they need to go.

Students generally select task cards in sequence because new concepts build on old ones in a specific order. There are, however, exceptions to this rule: students might *skip* a task that is not challenging; *repeat* a task with doubtful results; *add* a task of their own design to answer original "what would happen if" questions.

## 2. ORIENTATION

Many students will simply read a task card and immediately understand what to do. Others will require further verbal interpretation. Identify poor readers in your class. When they ask, "What does this mean?" they may be asking in reality, "Will you please read this card aloud?"

With such a diverse range of talent among students, how can you individualize activity and still hope to finish this module as a cohesive group? It's easy. By the time your most advanced students have completed all the task cards, including the enrichment series at the end, your slower students have at least completed the basic core curriculum. This core provides the common background so necessary for meaningful discussion, review and testing on a class basis.

## 3. INVESTIGATION

Students work through the task cards independently and cooperatively. They follow their own experimental strategies and help each other. You encourage this behavior by helping students only *after* they have tried to help themselves. As a resource person, you work to stay *out* of the center of attention, answering student questions rather than posing teacher questions.

When you need to speak to everyone at once, it is appropriate to interrupt individual task card activity and address the whole class, rather than repeat yourself over and over again. If you plan ahead, you'll find that most interruptions can fit into brief introductory remarks at the beginning of each new period.

## 4. WRITE-UP

Task cards ask students to explain the "how and why" of things. Write-ups are brief and to the point. Students may accelerate their pace through the task cards by writing these reports out of class.

Students may work alone or in cooperative lab groups. But each one must prepare an original write-up. These must be brought to the teacher for approval as soon as they are completed. Avoid dealing with too many write-ups near the end of the module, by enforcing this simple rule: each write-up must be approved *before* continuing on to the next task card.

## 5. CHECK POINT

The student and teacher evaluate each write-up together on a pass/no-pass basis. (Thus no time is wasted haggling over grades.) If the student has made reasonable effort consistent with individual ability, the write-up is checked off on a progress chart and included in the student's personal assignment folder or notebook kept on file in class.

Because the student is present when you evaluate, feedback is immediate and effective. A few seconds of this direct student-teacher interaction is surely more effective than 5 minutes worth of margin notes that students may or may not heed. Remember, you don't have to point out every error. Zero in on particulars. If reasonable effort has not been made, direct students to make specific improvements, and see you again for a follow-up check point.

A responsible lab assistant can double the amount of individual attention each student receives. If he or she is mature and respected by your students, have the assistant check the even-numbered write-ups while you check the odd ones. This will balance the work load and insure that all students receive equal treatment.

## 6. SCIENCE CONFERENCE

After individualized task card activity has ended, this is a time for students to come together, to discuss experimental results, to debate and draw conclusions. Slower students learn about the enrichment activities of faster students. Those who did original investigations, or made unusual discoveries, share this information with their peers, just like scientists at a real conference. This conference is open to films, newspaper articles and community speakers. It is a perfect time to consider the technological and social implications of the topic you are studying.

## 7. READ AND REVIEW

Does your school have an adopted science textbook? Do parts of your science syllabus still need to be covered? Now is the time to integrate other traditional science resources into your overall program. Your students already share a common background of hands-on lab work. With this shared base of experience, they can now read the text with greater understanding, think and problem-solve more successfully, communicate more effectively.

You might spend just a day on this step or an entire week. Finish with a review of key concepts in preparation for the final exam. Test questions in this module provide an excellent basis for discussion and study.

## 8. EXAM

Use any combination of the review/test questions, plus questions of your own, to determine how well students have mastered the concepts they've been learning. Those who finish your exam early might begin work on the first activity in the next new TOPS module.

Now that your class has completed a major TOPS learning cycle, it's time to start fresh with a brand new topic. Those who messed up and got behind don't need to stay there. Everyone begins the new topic on an equal footing. This frequent change of pace encourages your students to work hard, to enjoy what they learn, and thereby grow in scientific literacy.

# GETTING READY

Here is a checklist of things to think about and preparations to make before your first lesson.

## ☐ Decide if this TOPS module is the best one to teach next.

TOPS modules are flexible. They can generally be scheduled in any order to meet your own class needs. Some lessons within certain modules, however, do require basic math skills or a knowledge of fundamental laboratory techniques. Review the task cards in this module now if you are not yet familiar with them. Decide whether you should teach any of these other TOPS modules first: *Measuring Length, Graphing, Metric Measure, Weighing* or *Electricity* (before *Magnetism*). It may be that your students already possess these requisite skills or that you can compensate with extra class discussion or special assistance.

## ☐ Number your task card masters in pencil.

The small number printed in the lower right corner of each task card shows its position within the overall series. If this ordering fits your schedule, copy each number into the blank parentheses directly above it at the top of the card. Be sure to use pencil rather than ink. You may decide to revise, upgrade or rearrange these task cards next time you teach this module. To do this, write your own better ideas on blank 4 x 6 index cards, and renumber them into the task card sequence wherever they fit best. In this manner, your curriculum will adapt and grow as you do.

## ☐ Copy your task card masters.

You have our permission to reproduce these task cards, for as long as you teach, with only 1 restriction: please limit the distribution of copies you make to the students you personally teach. Encourage other teachers who want to use this module to purchase their *own* copy. This supports TOPS financially, enabling us to continue publishing new TOPS modules for you. For a full list of task card options, please turn to the first task card masters numbered "cards 1-2."

## ☐ Collect needed materials.

Please see the opposite page.

## ☐ Organize a way to track completed assignment.

Keep write-ups on file in class. If you lack a vertical file, a box with a brick will serve. File folders or notebooks both make suitable assignment organizers. Students will feel a sense of accomplishment as they see their file folders grow heavy, or their notebooks fill up, with completed assignments. Easy reference and convenient review are assured, since all papers remain in one place.

Ask students to staple a sheet of numbered graph paper to the inside front cover of their file folder or notebook. Use this paper to track each student's progress through the module. Simply initial the corresponding task card number as students turn in each assignment.

## ☐ Review safety procedures.

Most TOPS experiments are safe even for small children. Certain lessons, however, require heat from a candle flame or Bunsen burner. Others require students to handle sharp objects like scissors, straight pins and razor blades. These task cards should not be attempted by immature students unless they are closely supervised. You might choose instead to turn these experiments into teacher demonstrations.

Unusual hazards are noted in the teaching notes and task cards where appropriate. But the curriculum cannot anticipate irresponsible behavior or negligence. It is ultimately the teacher's responsibility to see that common sense safety rules are followed at all times. Begin with these basic safety rules:

1. Eye Protection: Wear safety goggles when heating liquids or solids to high temperatures.
2. Poisons: Never taste anything unless told to do so.
3. Fire: Keep loose hair or clothing away from flames. Point test tubes which are heating away from your face and your neighbor's.
4. Glass Tubing: Don't force through stoppers. (The teacher should fit glass tubes to stoppers in advance, using a lubricant.)
5. Gas: Light the match first, before turning on the gas.

## ☐ Communicate your grading expectations.

Whatever your philosophy of grading, your students need to understand the standards you expect and how they will be assessed. Here is a grading scheme that counts individual effort, attitude and overall achievement. We think these 3 components deserve equal weight:

1. Pace (effort): Tally the number of check points you have initialed on the graph paper attached to each student's file folder or science notebook. Low ability students should be able to keep pace with gifted students, since write-ups are evaluated relative to individual performance standards. Students with absences or those who tend to work at a slow pace may (or may not) choose to overcome this disadvantage by assigning themselves more homework out of class.

2. Participation (attitude): This is a subjective grade assigned to reflect each student's attitude and class behavior. Active participators who work to capacity receive high marks. Inactive onlookers, who waste time in class and copy the results of others, receive low marks.

3. Exam (achievement): Task cards point toward generalizations that provide a base for hypothesizing and predicting. A final test over the entire module determines whether students understand relevant theory and can apply it in a predictive way.

# Gathering Materials

Listed below is everything you'll need to teach this module. You already have many of these items. The rest are available from your supermarket, drugstore and hardware store. Laboratory supplies may be ordered through a science supply catalog.

Keep this classification key in mind as you review what's needed:

| | |
|---|---|
| *special in-a-box materials:*<br>Italic type suggests that these materials are unusual. Keep these specialty items in a separate box. After you finish teaching this module, label the box for storage and put it away, ready to use again the next time you teach this module. | general on-the-shelf materials:<br>Normal type suggests that these materials are common. Keep these basics on shelves or in drawers that are readily accessible to your students. The next TOPS module you teach will likely utilize many of these same materials. |
| *(substituted materials):*<br>Parentheses enclosing any item suggests a ready substitute. These alternatives may work just as well as the original, perhaps better. Don't be afraid to improvise, to make do with what you have. | *optional materials:<br>An asterisk sets these items apart. They are nice to have, but you can easily live without them. They are probably not worth an extra trip to the store, unless you are gathering other materials as well. |

Everything is listed in order of first use. Start gathering at the top of this list and work down. Ask students to bring recycled items from home. The teaching notes may occasionally suggest additional student activity under the heading "Extensions." Materials for these optional experiments are listed neither here nor in the teaching notes. Read the extension itself to find out what new materials, if any, are required.

Needed quantities depend on how many students you have, how you organize them into activity groups, and how you teach. Decide which of these 3 estimates best applies to you, then adjust quantities up or down as necessary:

$Q_1 / Q_2 / Q_3$

**Single Student:** Enough for 1 student to do all the experiments.
**Individualized Approach:** Enough for 30 students working alone or in small groups, all self-paced.
**Traditional Approach:** Enough for 30 students, working alone, all doing the same lesson.

| **KEY:** | *special in-a-box materials* | general on-the-shelf materials |
|---|---|---|
| | *(substituted materials)* | *optional materials |

| $Q_1 / Q_2 / Q_3$: | |
|---|---|
| 1/10/10 | rolls of pennies (paper punch — see activity 1) |
| 6/180/180 | sheets lined notebook paper |
| 1/30/30 | sheets copier paper |
| 1/15/30 | index cards |
| 1/15/30 | metric rulers |
| 1/15/30 | drawing compasses |
| 1/20/30 | scissors |
| 1/10/10 | rolls clear tape |
| 1/7/15 | hand calculators |
| 1/30/30 | *sheets dark construction paper* |
| 1/5/15 | bottles white glue |
| 1/5/15 | *hand lenses |
| 1/1/1 | pencil sharpener |
| 1/20/30 | *pieces 8 1/2 x 11 inch corrugated cardboard* |
| 1/2/2 | spools thread |
| 2/40/60 | straight pins |
| 1/20/30 | *dark crayons, wrappers removed* |
| 2/20/60 | *small styrofoam cups* |
| 1/5/15 | *short pieces pencil lead – see activity 12* |
| 4/120/120 | sheets scratch paper |
| 1/1/1 | roll masking tape |
| 1/2/2 | boxes paper clips |
| 1/1/1 | roll adding machine tape |
| 3/15/15 | *used envelopes with bar codes – see activity 19* |

# Sequencing Task Cards

This logic tree shows how all the task cards in this module tie together. In general, students begin at the bottom of the tree and work up through the related branches. As the diagram suggests, upper level activities build on lower level activities.

At the teacher's discretion, certain activities can be omitted, or sequences changed, to meet specific class needs. The only activities that must be completed in sequence are indicated by leaves that open *vertically* into the ones above them. In these cases the lower activity is a prerequisite to the upper.

When possible, students should complete the task cards in the same sequence as numbered. If time is short, however, or certain students need to catch up, you can use the logic tree to identify concept-related *horizontal* activities. Some of these might be omitted, since they serve only to reinforce learned concepts, rather than introduce new ones.

On the other hand, if students complete all the activities at a certain horizontal concept level, then experience difficulty at the next higher level, you might move back down the logic tree to have students repeat specific key activities for greater reinforcement.

For whatever reason, when you wish to make sequence changes, you'll find this logic tree a valuable reference. Parentheses in the upper right corner of each task card allow you total flexibility; they are left blank so you can pencil in sequence numbers of your own choosing.

## MATH LAB 07

E

# LONG-RANGE OBJECTIVES

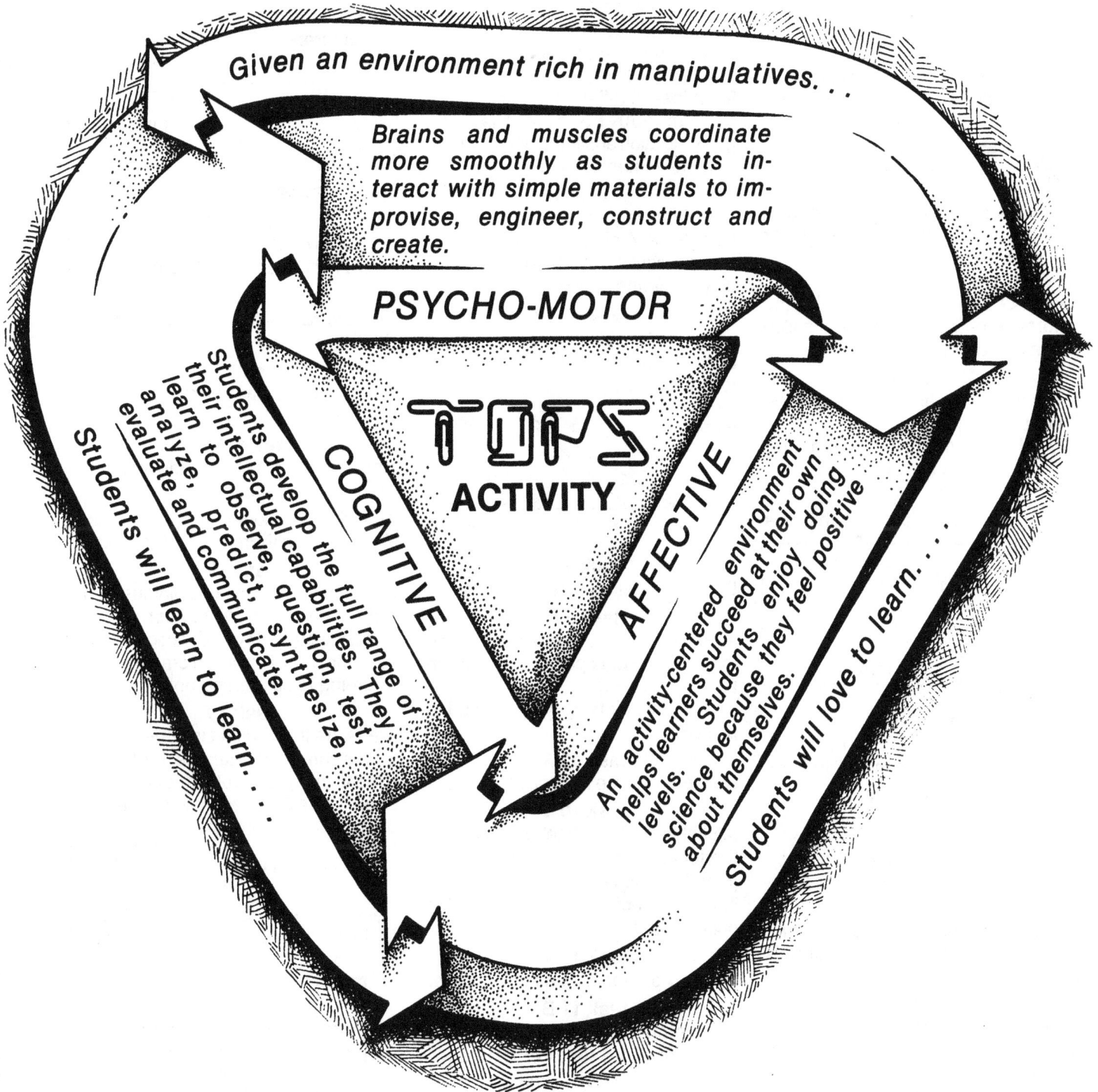

Given an environment rich in manipulatives. . .

Brains and muscles coordinate more smoothly as students interact with simple materials to improve, engineer, construct and create.

PSYCHO-MOTOR

TOPS ACTIVITY

COGNITIVE

Students develop the full range of their intellectual capabilities. They learn to observe, question, test, analyze, predict, synthesize, evaluate and communicate.

Students will learn to learn. . . .

AFFECTIVE

An activity-centered environment helps learners succeed at their own levels. Students enjoy doing science because they feel positive about themselves.

Students will love to learn. . . .

# Review / Test Questions

Photocopy this test question page. Cut out those questions you wish to use and tape them onto clean paper. Include questions of your own design, as well. Crowd them all onto a single page for students to answer on another paper, or leave space for student responses after each question, as you wish. Duplicate a class set and your custom-made test is ready to use. Use leftover questions as a class review in preparation for the final exam.

### tasks 1-2, 17 A
How many line segments are in the 6th term?

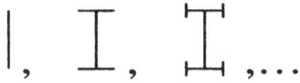

$$| \quad , \quad \mathsf{I} \quad , \quad \mathsf{I} \quad , \ldots$$

### task 1-2, 17 B
a. Draw the 4th term in this sequence with solid circles and hollow circles.

b. Write a general rule for n terms.
c. Use your nth term to find the number of circles in the 5th term.
d. Write out the first 5 terms as a sequence of numbers. Show that the difference between these terms also forms a sequence.

### task 3
a. Write out the first 16 terms in the Fibonacci sequence.
b. Add up these 16 terms the easy way. Explain your method.

### tasks 4-5 A
Use a calculator to show that this rectangle has "golden mean" proportions:

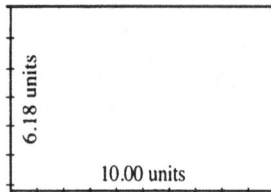

6.18 units

10.00 units

### tasks 4-5 B
Show how Fibonacci numbers approach the "golden mean" (0.618).

### tasks 6-7 A
You look under 47 rocks in a meadow and find ants under each one. You generalize that there will be ants under the next rock you find in this same meadow.
a. Is this a reasonable generalization?
b. Is it possible that your prediction is incorrect?

### tasks 6-7 B
a. Find the difference between adjacent terms in this sequence of prime numbers: 5, 7, 11, 13, 19, 23...
b. If this pattern holds, what should be the next prime number in this sequence?
c. What can you conclude?

### tasks 6-7 C
If both the length and width of a "pool table" have an odd number of squares, a cue ball hit at a 45° angle from 1 corner always exits the opposite corner.

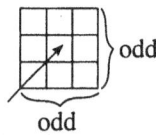

odd
odd

a. Show that this generalization is true for at least 5 different pool tables.
b. Can you be sure that this generalization applies to all pool tables with odd dimensions? Explain.

### task 8
A 3-sided regular polygon forms a triangle. A 4-sided regular polygon forms a square. What sort of regular polygon forms a circle?

### tasks 8-12
These 4 straight lines approximate a curve. Draw more lines to approximate it even better.

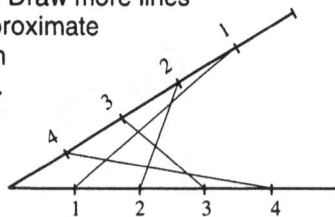

### task 9 A
Define a circle as a set of points.

### task 9 B
a. Sketch a circle freehand, marking the center c, plus points $p_1$ and $p_2$ on the circle.
b. How are points $p_1$ and $p_2$ related to point c?

### task 10 A
Define an ellipse as a set of points.

### task 10 B
a. Sketch an ellipse freehand, marking the foci $f_1$ and $f_2$, plus points $p_1$ and $p_2$ on the ellipse.
b. How are points $p_1$ and $p_2$ related to points $f_1$ and $f_2$?

### task 11 A
Define a parabola as a set of points.

### task 11 B
a. Sketch a parabola freehand, marking its focus f and fixed reference line l, plus points $p_1$ and $p_2$ on the parabola.
b. How are points $p_1$ and $p_2$ related to the focus f and line l?

### task 12 A
Define a cardioid as a set of points.

### task 12 B
A movable circle is rotated about a fixed circle as shown. What kind of figure is traced by point A on the circumference? Point B at the center?

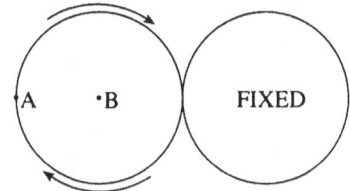

A  ·B    FIXED

### task 13
A triangle has 3 equal interior angles. How big is each one? Explain.

### task 14
Find the sum of the first 200 integers. Show your work.

### task 15 A
Find 2 pairs of twin primes between 100 and 110.

### task 15 B
Show that 143 is not a prime number.

### task 16
To multiply numbers, _____ their logs.
To divide numbers, _____ their logs.

### task 18
a. Explain how to make a Mobius strip from a 1 meter strip of paper.
b. If an ant travels in 1 direction along this strip, how far must it crawl before returning to its starting point?

### task 19
Translate this 5-digit zip code. Show that its correction character is correct. (Recall that place values for the long bars are 7 4 2 1 0, and that the bar on each end is a frame bar.)

### task 20
Here is 1 way to connect 4 tiles. Draw all the other ways. Each pattern must be unique.

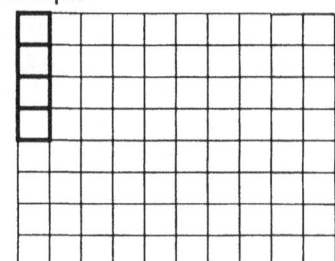

# Answers

### tasks 1-2, 17 A

1, 3, 7, 15, 31, <u>63</u> — line segments

add: 2   4   8   16   32

### tasks 1-2, 17 B

a. (filled/open circle grid)

b. $n^2 + n$

c. $5^2 + 5 = 30$

d. 2, 6, 12, 20, 30, ... total circles

    4, 6, 8, 10, ... differences

### task 3

a. 1, 1, 2, 3, 5, 8, 13, 21, 34, 55, 89, 144, 233, 377, 610, 987,...

b. The sum of the first 14 terms is one less than the 16th term (986). The sum of this entire Fibonacci sequence is thus: 986 + 610 + 987 = 2,583

### task 4-5 A

Golden rectangles have sides with these special proportions:

$$\frac{w}{l} = \frac{l}{w+l}$$

$$\frac{6.18}{10.00} = \frac{10.00}{10.00 + 6.18} = 0.618$$

### task 4-5 B

The Fibonacci sequence is:
1, 1, 2, 3, 5, 8, 13, 21, 34, 55, 89, ....
The ratio of adjacent terms in this sequence approximates the golden mean as the numbers increase:

$$\frac{3}{5} = 0.600, \quad \frac{8}{13} = 0.615, \quad \frac{21}{34} = 0.618,$$

### tasks 6-7 A

a. Yes.

b. Yes. Generalizations are not for sure. An exception to the pattern may yet be discovered.

### tasks 6-7 B

a. 5, 7, 11, 13, 17, 19, 23,...

   2  4  2  4  2  4

b. Extending this pattern, the next prime term should be: 23 + 2 = 25.

c. Twenty-five is not a prime number. This teaches the important lesson that a good generalization does not necessarily lead to a correct prediction.

### tasks 6-7 C

a.

b. No. We haven't looked at all odd pool tables. An exception to this rule may yet be discovered.

### task 8

A circle is a regular polygon of n sides, where n is a very large number.

### tasks 8-12

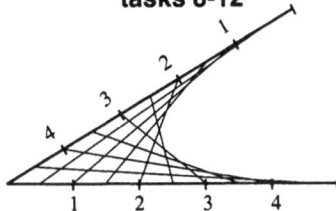

### task 9 A

A circle is the set of all points in a plane that are the same distance from a fixed point (the center).

### task 9 B

a.

b. Points $p_1$ and $p_2$ are both equidistant to point c.

### task 10 A

An ellipse is the set of all points in a plane such that the sum of the distance of each point from two fixed points (the foci) is the same.

### task 10 B

a.

b. The sum of the distances from each point to $f_1$ and $f_2$ are equal:

$$\overline{p_1f_1} + \overline{p_1f_2} = \overline{p_2f_1} + \overline{p_2f_2}$$

### task 11 A

A parabola is the set of all points in a plane such that the distance of each point to a fixed point (the focus) is the same as its distance to a fixed line.

### task 11 B

a.

b. Point $p_1$ is equidistant from both the focus f and line l. The same is true for point $p_2$.

### task 12 A

A cardioid is the set of points traced by a point on the circumference of a circle rolling completely around a fixed circle of equal size.

### task 12 B

Point A traces a cardioid. Point B traces a circle.

### task 13

The sum of the interior angles of a triangle is 180°. If all 3 angles are equal, each angle has a value of 180°/ 3 = 60°.

### task 14

The first 200 integers may be paired like these:

200 199 198 102 ..., 101
  1,  2,  3, 99 ..., 100

There are 100 such pairs, each with the sum of 201:

100 x 201 = 20,100.

### task 15 A

101 and 103, 107 and 109.

### task 15 B

143 = 11 x 13.

### task 16

<u>add</u>, <u>subtract</u>

### task 18

a. Give the strip of paper a half twist, then join it end to end.

b. The ant must travel 2 meters in the same direction to return to its starting point.

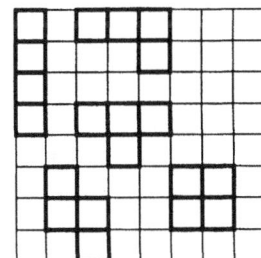

9  2  0  8  3  8

### task 19

9 + 2 + 0 + 8 + 3 + 8 = 30.

Thirty is multiple of 10.

### task 20

There are 5 unique ways to group 4 squares.

# TEACHING NOTES

## For Activities 1-20

**Task Objective (TO)** discover patterns in the arrangement of pennies that form number sequences.

**PENNY PLAY**   O                                    Math Lab (    )

1. These penny patterns form number sequences. Write the first 10 terms of each sequence as a string of numbers.

a. O  ...    (The differences between neighboring terms also forms a number pattern.)
   1,   3,   6,

b. O  ...    (These numbers are called *perfect squares*.)
   1,   4,   9,

c. O  ...   (Build the next term like this: 1 "heads" surrounded by 6 tails, surrounded by 12 heads, surrounded by more tails...)
   1,   7,   19,

2. Use the sequences you have just written to help you find the 10th term in each of these sequences:

   a. 1, 1+2, 1+2+3, ...
   b. 1, 1+3, 1+3+5, ...
   c. 1, 1+7, 1+7+19, ...

3. Here is a penny puzzle: This triangle points up. Move just 3 pennies to make it point down.

1

## Answers / Notes

1a. 1, 3, 6, 10, 15, 21, 28, 36, 45, 55. *(The difference between pairs of terms is 2, 3, 4, ....)*

1b. 1, 4, 9, 16, 25, 36, 49, 64, 81, 100. *(The difference between pairs of terms is 3, 5, 7, ....)*

1c. 1, 7, 19, 37, 61, 91, 127, 169, 217, 271. *(The difference between pairs of terms is 6, 12, 18, .... This corresponds to the number of pennies added in each new outside ring.)*

2a. This sequence is the same as 1a. The tenth term, therefore, is 55.

2b. This sequence is the same as 1b. The tenth term, therefore, is 100.

2c. The 10th term in this sequence equals the sum of all 10 terms in 1c: $1 + 7 + 19 + 37 + 61 + 91 + 127 + 169 + 217 + 271 = 1000$ *(Students who recognize problem 2c as a sequence of perfect cubes can arrive at the same answer by cubing the tenth term: $10^3 = 1000$.)*

3. *The key to solving this puzzle is recognizing the triangle as a 7-penny hexagon surrounded by 3 outside pennies. Move these 3 to point the triangle in any of a total of 6 different directions, including down.*

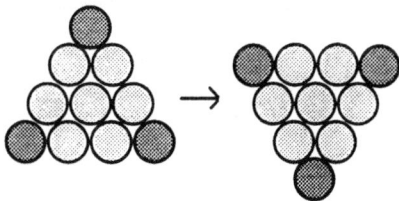

## Materials

☐ A roll of pennies. Problem 1c requires 37 coins to build the 4th term. If you don't want to invest several dollars to supply the needs of your whole class, substitute index card punches. Color 1 side before punching so each side has a "head" and a "tail." The only requirement for using these is good eye-hand coordination. Use a straight pin to slide them into position.

**(TO)** discover patterns in the arrangement of squares that form number sequences. To write a general term for each sequence.

---

## GENERAL TERM ○ Math Lab ( )

For each sequence below…

1. Draw the 4th term, showing 4 shaded squares.

2. Write a general term (**n**th term) for **n** shaded squares.

3. Use this general term to find the 5th term.

4. Write the first 5 terms as numbers. Check that the differences between terms also forms a pattern.

a.

b.

c.

d.

e.

f.

2

---

## Introduction

List these number sequences on your blackboard. Express the two bottom sequences as patterns of squares.

| Term: | 1st | 2nd | 3rd | 4th | 5th | …nth |
|-------|-----|-----|-----|-----|-----|------|
| | 1 | 2 | 3 | 4 | 5 | $n$ |
| | 2 | 4 | 6 | 8 | 10 | $2n$ |
| | 1 | 3 | 5 | 7 | 9 | $2n-1$ |
| | 1 | 4 | 9 | 16 | 25 | $n^2$ |
| | 0 | 2 | 6 | 12 | 20 | $n^2-n$ |

ANY NUMBER

## Answers / Notes

1a. 1b. 1c. 1d. 1e. 1f.

| 2-4. | General term: | 5th term: | first 5 terms: | differences: |
|------|---------------|-----------|----------------|--------------|
| a. | $3n$ | 15 | 3, 6, 9, 12, 15 | 3, 3, 3, 3 |
| b. | $n^2$ | 25 | 1, 4, 9, 16, 25 | 3, 5, 7, 9 |
| c. | $3n+1$ | 16 | 4, 7, 10, 13, 16 | 3, 3, 3, 3 |
| d. | $n^2 + n$ | 30 | 2, 6, 12, 20, 30 | 4, 6, 8, 10 |
| e. | $n^2 + n + 1$ | 31 | 3, 7, 13, 21, 31 | 4, 6, 8, 10 |
| f. | $2n^2 - 1$ | 49 | 1, 7, 17, 31, 49 | 6, 10, 14, 18 |

## Materials

☐ None.

---

## THE FIBONACCI SEQUENCE  ○                     Math Lab (    )

a.

$$1, 1, 2, 3, 5, 8, 13, \ldots$$

b.

1. The Italian mathematician Leonardo Fibonacci first investigated this sequence of numbers in 1683.

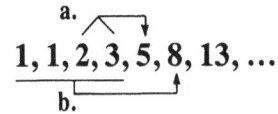

   a. How are the previous 2 terms always related to the next term? Use your rule to extend this sequence to 15 terms.

   b. How are the *sum of the first n terms* in this sequence related to the number that follows 2 terms later?

2. An alien life form, a juvenile **b**lob, reaches Earth in a space pod. In 1 day it grows to an adult **B**lob (as big as a pea). In 1 more day (and every day thereafter) this adult Blob gives birth to another juvenile blob, each of which matures and reproduces like its parent.

FEED
ME!

   a. Track the blob population for *1 week* on lined paper, like this. ⟶

   b. What is the total blob population after the 30th day?

   c. Each blob consumes 1 gram of biomass per day. How much does this alien life form consume after 30 days? (1,000,000 grams = 1 metric ton)

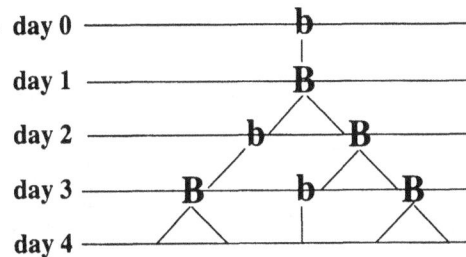

| | |
|---|---|
| day 0 | b |
| day 1 | B |
| day 2 | b        B |
| day 3 | B      b      B |
| day 4 | |

© 1994 by TOPS Learning Systems                     3

## Answers / Notes

2a.

| DAY | POPULATION |
|---|---|
| 0 | 1 |
| 1 | 1 |
| 2 | 2 |
| 3 | 3 |
| 4 | 5 |
| 5 | 8 |
| 6 | 13 |
| 7 | 21 |
| 8 | 34 |
| 9 | 55 |
| 10 | 89 |
| 11 | 144 |
| 12 | 233 |
| 13 | 377 |
| 14 | 610 |
| 15 | 987 |
| 16 | 1,597 |
| 17 | 2,584 |
| 18 | 4,181 |
| 19 | 6,765 |
| 20 | 10,946 |
| 21 | 17,711 |
| 22 | 28,657 |
| 23 | 46,368 |
| 24 | 75,025 |
| 25 | 121,393 |
| 26 | 196,418 |
| 27 | 317,811 |
| 28 | 514,229 |
| 29 | 832,040 |
| 30 | 1,346,269 |

1a. The previous 2 terms always add up to the next term. Applying this rule to the first 15 terms in the Fibonacci sequence yields: 1, 1, 2, 3, 5, 8, 13, 21, 34, 55, 89, 144, 233, 377, 610.

1b. The first n terms add up to 1 less than the number 2 terms later. Example: $1 + 1 + 2 + 3 = 8 - 1$.

2b. The daily population follows the Fibonacci sequence. At the end of day 30 it is 1,346,269 blobs.

2c. Total food consumption in grams equals the sum of this Fibonacci sequence through day 30:

    days 0-28  =  1,346,268 *(This is the population of day 30 less 1)*

    day 29  =     832,040

    day 30  =  1,346,269

    total  =  3,524,577 grams consumed

    ≈ 3.5 metric tons!

## Materials

☐ Lined notebook paper.

**(TO)** draw Fibonacci squares and construct the spiral of a chambered nautilus.

## SEASHELL SPIRAL

○

1. Draw a 14.4 cm square on paper with square corners. Use an index card to make right angles.

   a. Inscribe a quarter circle from corner to corner with a compass.
   b. Cut out the square.

2. Draw, inscribe, and cut smaller squares of 8.9 cm, 5.5 cm, 3.4 cm, 2.1 cm, and 1.3 cm from your paper. Tape them together so the curved lines spiral inward.

8.9 cm
TAPE ON BACK
14.4 cm
5.5 cm

3. Tape paper *behind* the hole that remains. Draw and inscribe smaller squares so the line spirals to a point.

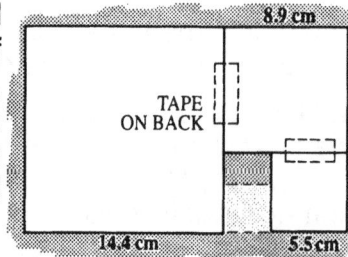

PATCH HOLE FROM BACK

4. These squares define the shape of a chambered nautilus. What determines the size of each square?

5. Where else can you find Fibonacci numbers?

COSMOS
PERIWINKLE
GERANIUM LEAF
PIANO KEYS

© 1994 by TOPS Learning Systems

4

## Answers / Notes

1-3. *The taped squares with inscribed circles spiral together like these (1/2 actual size). The 6 largest squares are cut and taped. Students should draw the smaller inner squares freehand.*

4. The dimensions of each square come from the Fibonacci sequence.

5. The 5 black keys and 8 white keys in a piano octave are Fibonacci numbers. These numbers are also common in the petals of flowers and the lobes or segments of leaves, as in those illustrated.

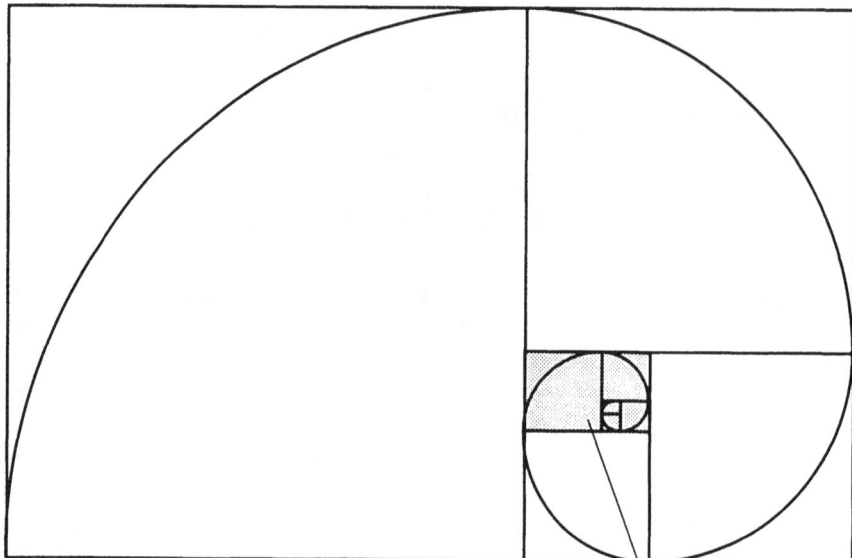

Students draw this part on paper patch.

## Materials

☐ Paper with square corners. Use copier paper if your notebook paper has rounded corners.
☐ An index card.
☐ A metric ruler.
☐ A drawing compass.
☐ Scissors.
☐ Clear tape.

**(TO)** study the golden mean as a ratio of Fibonacci numbers. To create a work of art based on this proportion.

---

## THE GOLDEN MEAN ○           Math Lab ( )

1. Measure the length (**L**) and width (**W**) of all 8 <u>Golden</u> <u>Rectangles</u> to the nearest whole mm. Use a calculator (to 3 decimal places) to complete this table.

| L(mm) | W(mm) | L+W | $\frac{W}{L}$ | $\frac{L}{L+W}$ |
|-------|-------|-----|-----|------|
| 233 | 144 | 377 | .618 | .618 |

↓   (List largest to smallest.)  ↓

    a. Do you recognize the numbers in the first 3 columns of your table?
    b. The last 2 columns contain ratios that equal the *golden mean*. Write a definition of the golden mean.
    c. If you were to cut a square off a golden rectangle, what remains?

2. Carefully cut around the outer rectangle. Fold it in half so all edges match.
    a. Snip into the folded borders of all 7 rectangles.
    b. Unfold, then finish cutting out all 7 "picture frames."

SNIP TO START

*PASTE DOWN A DESIGN YOU LIKE!*

3. The golden mean has been used in art and architecture since ancient times. Arrange your "golden" frames on dark construction paper to create your own eye-catching design.

© 1994 by TOPS Learning Systems

5

---

### Answers / Notes

1. Students should complete the above table with 8 lines.
1a. The first 3 columns contain Fibonacci numbers.
1b. The golden mean, 0.618, is the ratio you get when dividing the width of a golden rectangle by its length, and when dividing its length by the sum of its width and length.
1c. Cut a square off any golden rectangle, and you get a smaller golden rectangle. For example, if you remove a 144 x 144 mm square from a 233 x 144 mm rectangle, you are left with a smaller 144 x 89 mm rectangle that again has golden mean proportions.
3. *Your students might work with whole frames or with parts. Encourage them to experiment with many different arrangements before gluing a final design. Here are two of many possibilities:*

### Materials

☐ The Golden Rectangles cutout. Photocopy this from the supplementary pages at the back of this book.
☐ A metric ruler.
☐ A calculator.
☐ Scissors.
☐ Dark construction paper.
☐ White glue.

**(TO)** generate a sequence of regions by joining points on the circumference of a circle. To recognize the tentative nature of predictions based on generalizations.

---

### CIRCLE SEQUENCE ○ Math Lab ( )

1. All points on these circles are joined to all other points by straight lines. This makes a sequence of increasing points and regions.

| 1 POINT | 2 POINTS | 3 POINTS | 4 POINTS | 5 POINTS | 6 POINTS |
|---|---|---|---|---|---|
| 1 REGION | 2 REGIONS | 4 REGIONS | 8 REGIONS | ? REGIONS | ? REGIONS |

a. Predict the number of regions in the 5th and 6th terms.
b. On what pattern did you base your generalization?

2. Use a compass to draw large copies of the circles with 5 and 6 points equally spaced around the outside. Draw straight lines joining every point with *every* other point in the circle.

3. Count the regions in each circle.
a. Was your <u>prediction</u> correct? Explain.
b. Was your <u>generalization</u> reasonable? Explain.
c. What does this activity tell you about predictions and generalizations?

© 1994 by TOPS Learning Systems                                                    6

---

### Answers / Notes

1a. Prediction: The 5th term will have 16 spaces; the 6th term will have 32 spaces: 1, 2, 4, 8, <u>16</u>, <u>32</u>.

1b. Each new term is twice the previous term. *(This can also be written as a sequence of base 2 exponents: $2^0$, $2^1$, $2^2$, $2^3$, $2^4$, $2^5$.)*

2-3. *Watch out for these errors as students draw their circles:*
*• Incomplete drawings: every point on the 5 point circle must radiate 4 straight lines; every point on the 6 point circle must radiate 5 straight lines.*
*• Counting the same space twice or not at all: check each region as it is counted until all regions are marked.*

The 5 point circle has 16 regions as predicted.

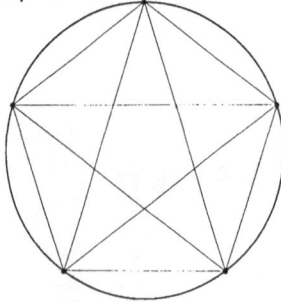

The 6 point circle has 30 regions, two less than predicted!

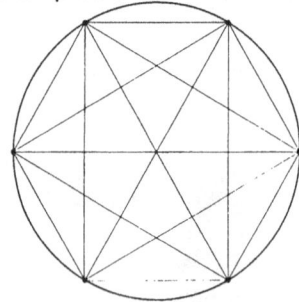

*If points on the circumference are not equally spaced, an additional region opens at the center of the circle where 3 lines now cross at the same point. These circles will have 31 regions, 1 less than the predicted 32.*

3a. The prediction that a 5 point circle should have 16 regions was correct. The prediction that a 6 point circle should have 32 regions was too large. There were only 30 regions.

3b. Yes. The sequence 1, 2, 4, 8… does double with each term. It was reasonable to generalize that this pattern would continue.

3c. A good generalization does not necessarily insure a correct prediction. An exception to the ruler may be hiding just out of sight, as it was in the 6th term.

### Extension

Find how many regions are in a 7 point circle. (57)

### Materials

☐ A drawing compass.

---

**(TO)** trace the path of a ball as it is deflected across a hypothetical pool table. To recognize the tentative nature of predictions based on generalizations.

---

## POOL TABLE INDUCTION ○                              Math Lab (   )

1. Copy these 6 "pool tables" on a <u>Centimeter</u> <u>Grid</u>.

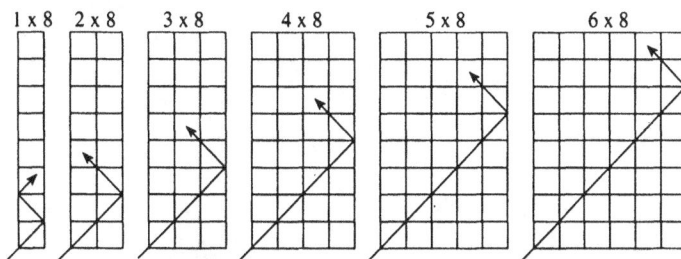

1 x 8   2 x 8   3 x 8   4 x 8   5 x 8   6 x 8

*EVERY BOUNCE FORMS A RIGHT ANGLE.*

2. A cue ball is hit from the lower left corner of each table and bounces as shown. Draw the ball's path across each table until it finally drops into a corner pocket. (Assume there are no side pockets.)
  a. In which pocket does the cue ball always drop?
  b. If this trend continues, where should the cue ball drop on a 7x8 and 8x8 pool table? Make a generalization.
  c. Is this generalization certain?

3. Test your prediction.
  a. Draw 7x8 and 8x8 pool tables on graph paper and "shoot" the cue ball. What happens?
  b. What have you learned about making generalizations?

© 1994 by TOPS Learning Systems                              7

---

## Answers / Notes

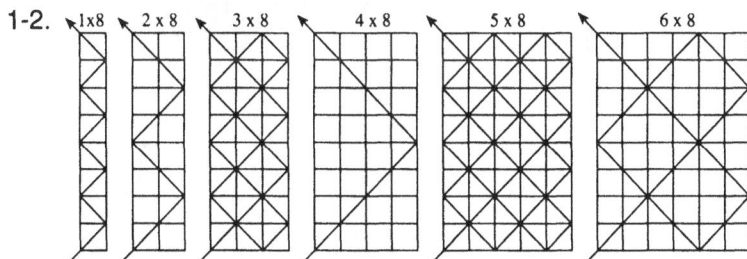

1-2.  1x8   2 x 8   3 x 8   4 x 8   5 x 8   6 x 8

2a. The cue ball drops into the upper left pocket of each pool table.
2b. Generalization: If this trend continues, the cue ball will continue to drop into the upper left pocket of each pool table.
2c. This generalization is not certain. An exception to the rule, yet unseen, may exist.

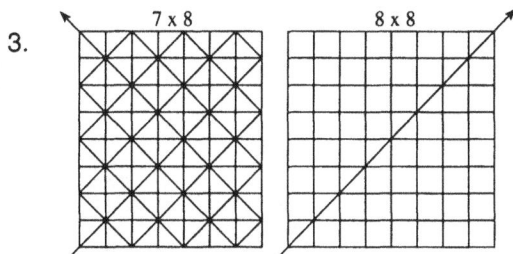

3.   7 x 8   8 x 8

3a. In a 7x8 table, the cue ball drops in the upper left pocket as predicted.
   In an 8x8 table, the cue ball drops directly into the upper right pocket without bouncing at all!

3b. Generalizing is risky business. An exception to the rule may be hiding just out of sight.

## Extension

Study pool tables with different grid dimensions. Can you generalize a rule that successfully predicts when a pool table "fills up" with bounces and when it doesn't?
*A pool table "fills up" when its length and width have no common factors.*

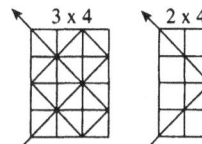

3 x 4   2 x 4

## Materials

☐ The Centimeter Grid. Photocopy this from the supplementary pages.

---

**(TO)** approximate a curve with straight lines. To recognize that n-sided polygons approach the shape of a perfect circle as n increases.

## TO THE LIMIT ◯ Math Lab ( )

1. Cut out the <u>Number Square</u>. There are 4 pairs of points in section A labeled 1, 2, 3 and 4. Join each pair with a straight line.

2. Join pairs of points in sections B, C and D in a similar manner with a sharpened pencil.

3. Examine the perimeter of the central shape within each section A, B, C and D. Do you see…
- straight lines crossing at angles? (or)
- a smooth curved line without angles?

*HOW SMOOTH IS THIS PERIMETER IN EACH QUADRANT?*

4. Consider this sequence of regular polygons.

△ , □ , ⬠ , ⬡ , ⬡ , ⬡ , ⬡ , ⬡ ,…

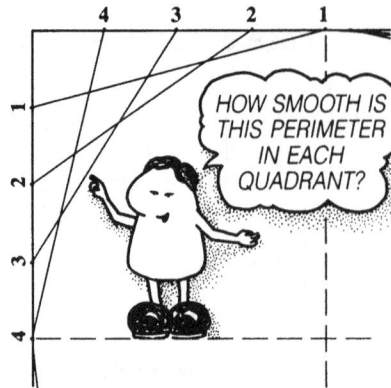

a. As the number of sides grow without limit, what shape does the polygon eventually assume?

b. How many straight sides are required to form a perfect circle? How long is each side?

8

## Answers / Notes

1-2.

3A. Straight lines crossing at angles.

3B. Straight lines crossing at angles.

3C. Straight lines crossing at angles. (or) A smooth curved line without angles. *(This is a transition zone. Angles between line segments may or may not be apparent depending on pencil sharpness and magnification.)*

3D. A smooth curved line without angles.

4a. The polygon eventually assumes the shape of a circle.

4b. A circle has an infinite number of straight sides that are infinitely short.

## Materials

☐ A hand lens (optional).
☐ The Number Square supplement.
☐ A sharp pencil.
☐ Scissors.
☐ A ruler or straight edge.

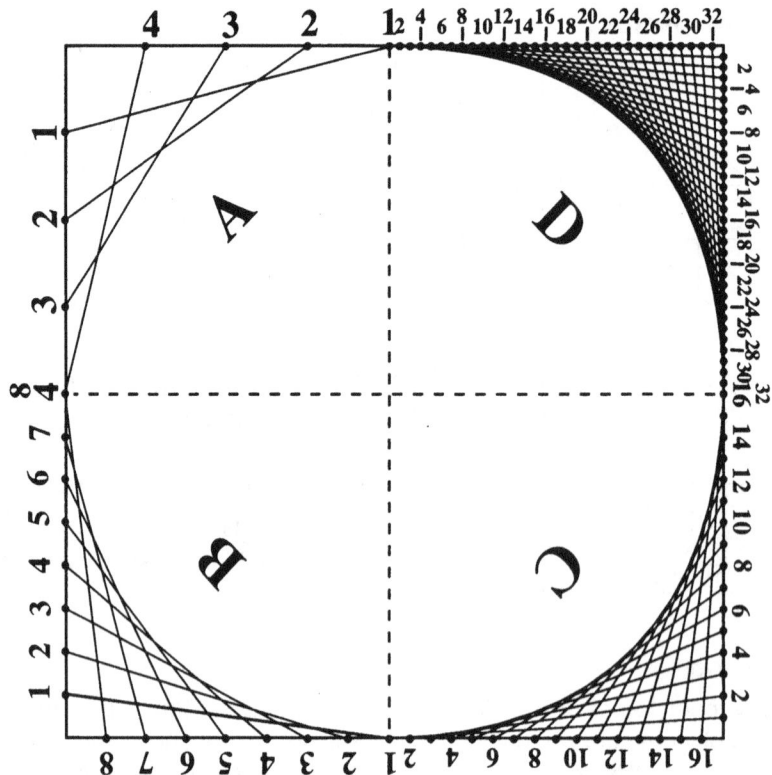

**(TO)** define a circle and draw it with a smooth curve. To approximate another circle with many straight fold lines.

---

### CIRCLE                                        O                              Math Lab (    )

1. Cut corrugated cardboard to notebook paper size.

2. Cut a piece of thread equal to the width of notebook paper. Tie it in a loop.

3. Center the paper on the cardboard and stick a pin in the middle. Put the loop over the pin.

4. Tighten the loop with your pencil point, and draw a circle around the pin. Mark the center "c" and points $p_1$, $p_2$, $p_3$, $p_4$ like these:

5. Copy this definition: A *circle* is the set of all points in a plane that are the same distance from a fixed point (the center).

   a. What plane do these 5 points share?

   b. Which points "p" are part of the circle? Why?

6. Cut out your circle. Fold over the edge to touch the center, then crease sharply with your fingernail. Do this repeatedly until you define a new shape. What is it?

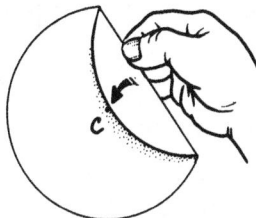

7. Turn the circle over so the creases form raised ridges. Accent the ridges with the *side* of a dark crayon.

© 1994 by TOPS Learning Systems                                                                    9

---

### Answers / Notes

5. A circle is the set of all points in a plane that are the same distance from a fixed point (the center).

5a. All 5 points (c, $p_1$, $p_2$, $p_3$, and $p_4$) share the same plane as the notebook paper.

5b. Points $p_1$, $p_2$ and $p_3$ are all part of the circle because they are all the same distance from the center point c. Point $p_4$ is not part of this circle because it is not equidistant. It lies closer to c than the other 3 points.

6. The new shape is a smaller circle. *(It has half the radius of the cutout circle.)*

7. The new half-sized circle now stands out in bolder relief.

   *Students should save the corrugated cardboard and thread loop to use again.*

### Materials

- ☐ Corrugated cardboard.
- ☐ Scissors.
- ☐ Thread.
- ☐ Notebook paper.
- ☐ A straight pin.
- ☐ A dark crayon with the paper wrapper removed.

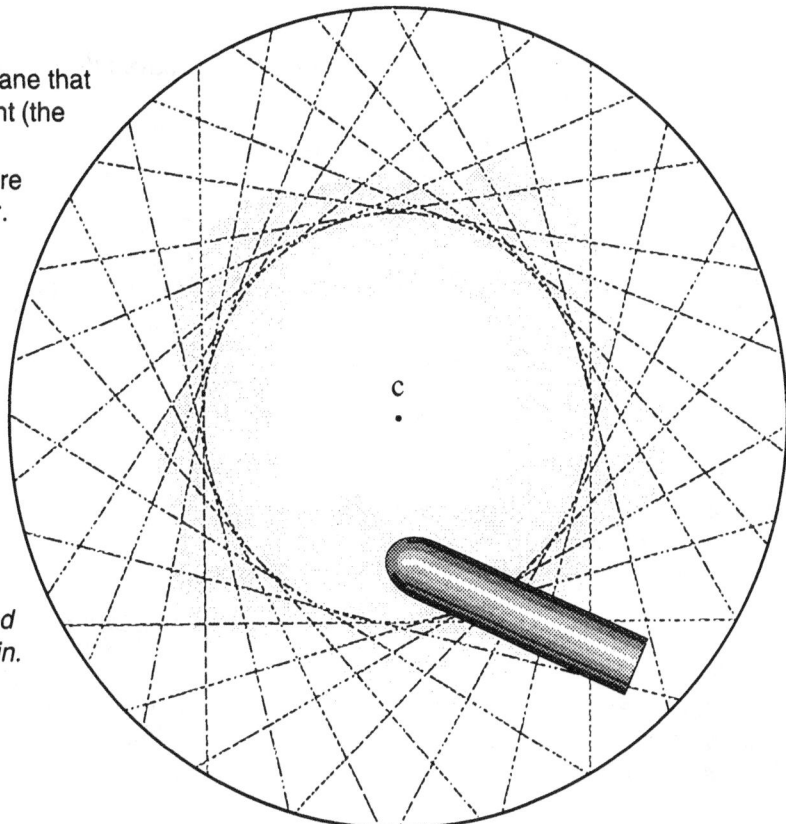

**(TO)** define and construct an ellipse. To form a curved figure from straight fold lines and evaluate its shape.

**ELLIPSE**   O   **Math Lab (   )**

1. Set notebook paper on cardboard as before. Lay your thread loop in a circle centered on the paper.

2. Stick 2 pins just *inside* the loop at opposite sides of the circle.

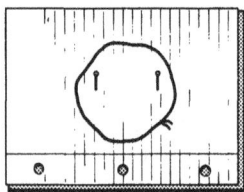

3. Tighten the loop with your pencil point and draw around both pins. Mark the foci $f_1$ and $f_2$ at each pin, and points $p_1$, $p_2$ and $p_3$ like these:

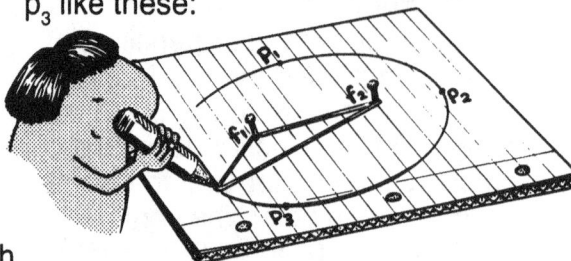

4. Copy this definition: An *ellipse* is the set of all points in a plane such that the sum of the distance of each point from two fixed points (called the foci) is the same. Measure to each point "p" to show this is true.

5. Cut out your ellipse. Fold the edge in to touch focus $f_1$ (not $f_2$), then crease the fold. Repeat over and over until you define a new shape.

6. Darken the ridges on the back with crayon as before. Describe the shape that emerges.

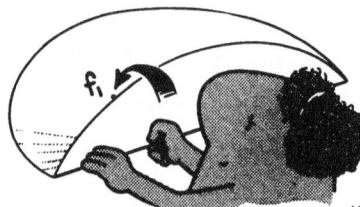

© 1994 by TOPS Learning Systems

10

## Answers / Notes

1. *Students should make a new thread loop if they haven't saved the old one. As before, the thread should be cut to the width of notebook paper.*

4. An ellipse is the set of all points in a plane such that the sum of the distance of each point from two fixed points (called the foci) is the same.

*Students should verify this with a metric ruler. These measurements are based on the example at right.*

$\overline{p_1f_1} + \overline{p_1f_2} = 48.0$ mm $+ 48.5$ mm $= 96.5$ mm

$\overline{p_2f_1} + \overline{p_2f_2} = 72.0$ mm $+ 24.0$ mm $= 96.0$ mm

$\overline{p_3f_1} + \overline{p_3f_2} = 30.0$ mm $+ 66.5$ mm $= 96.5$ mm

6. The straight fold lines define an egg shape that is more rounded on one side and more pointed on the other. It is not an ellipse.

## Materials

☐ Corrugated cardboard and the thread loop from the previous activity.
☐ Notebook paper.
☐ Two straight pins.
☐ A metric ruler.
☐ Scissors.
☐ A dark crayon with paper wrapper removed.

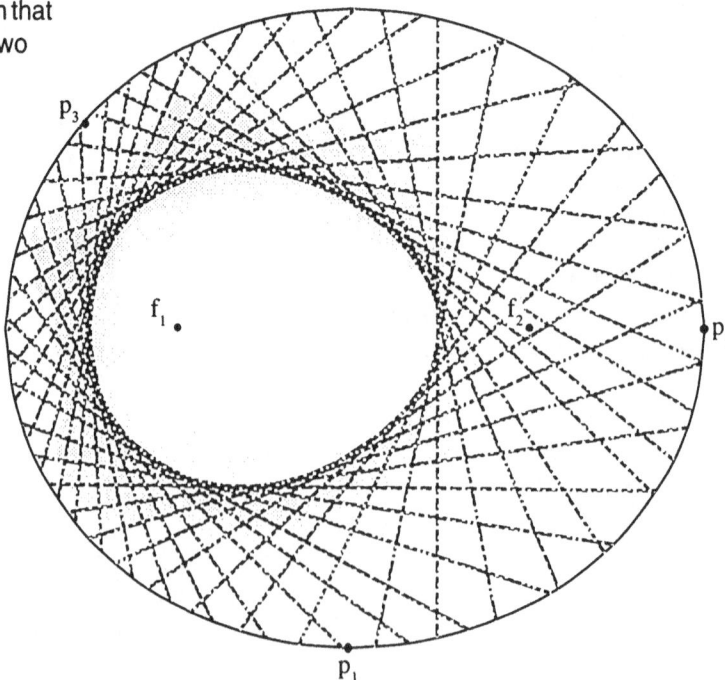

**(TO)** construct a parabola using many straight fold lines. To develop a definition based on its properties.

---

## PARABOLA      ◯      Math Lab ( )

1. Carefully trim notebook paper along its margin line. Center a pinhole 4 line widths inside this margin. Label it "f" for focus.

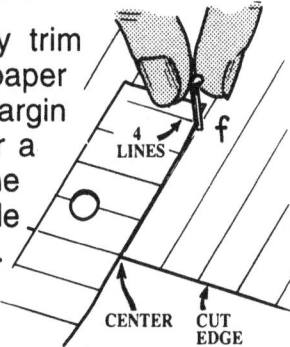

2. Fold over this margin edge to touch the focus point, then crease the fold. Repeat all along the margin until the creases define a *parabola*. Shade the ridges on the back with crayon as before.

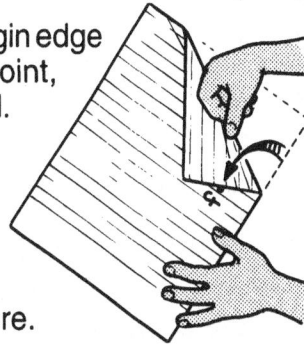

3. Tie thread around a pin. Leave one end as long as your paper. Trim off the other end.

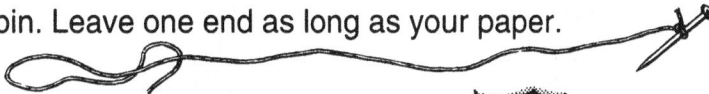

4. Poke the pin into your curved *parabola* at different points.

    a. At each point, use thread to compare distances from pinhead to focus; from pinhead to margin.

    b. Finish this definition: A *parabola* is the set of all points in a plane such that...

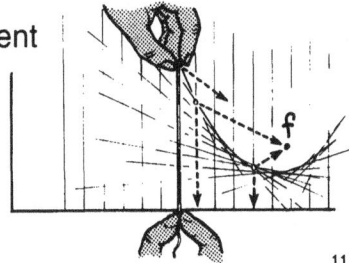

© 1994 by TOPS Learning Systems      11

---

## Answers / Notes

1. Students should use the cut-off margin strip as a convenient "ruler" to measure 4 line widths. The pinhole should be centered lengthwise on the paper.

4a. The same length of thread reaches both the focus and the margin from any point on the parabola. *(The distance from pin to margin is measured parallel to the lines on the notebook paper, perpendicular to the margin. This is the shortest distance from the pin to the margin.)*

4b. A parabola is the set of all points in a plane such that the distance of each point to a fixed point (the focus) is the same as its distance to a fixed line.

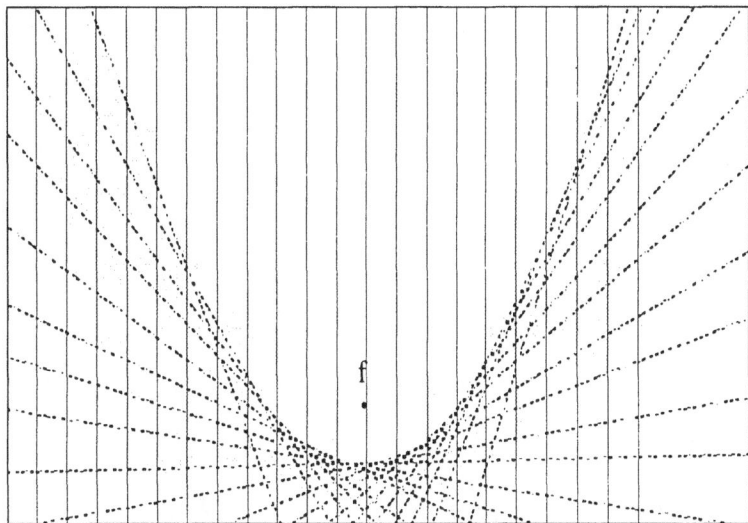

f

## Materials

☐ Notebook paper.
☐ Scissors.
☐ A straight pin.
☐ Thread.
☐ A dark crayon.

(TO) define a cardioid and draw it using circles of Styrofoam. To approximate another cardioid with straight lines.

## CARDIOID ⭕ Math Lab ( )

CUT SIDES AWAY | TURN OVER

SMOOTH EDGES

1. Cut away the sides of two styrofoam cups. Smooth the upper surfaces of the remaining disks. Turn these so the rims face up.

2. Poke a short piece of pencil lead through the rim of one disk so it sticks out about 1 mm beyond the flat side. Glue it in place on the rim.

GLUE
RIM
DISK
LEAD

PAPER
ROLLING DISK
PENCIL LINE
GLUED DISK
PENCIL LEAD
Roll, don't slide.

3. Lightly glue the flat side of the other disk to the center of some paper.

4. Roll the free disk all the way around the fixed disk so the pencil lead leaves a smooth line. Roll the disks against each other — don't let them slide.
   a. Describe the *cardioid* that you have drawn.
   b. Write a definition for this term.

5. Get a set of <u>Cardioid</u> Points. Connect them with straight lines so each point is joined to a number that is twice its value: 1 to 2, 2 to 4, 3 to 6, 4 to 8,....

© 1994 by TOPS Learning Systems

12

## Answers / Notes

2. Position the pencil lead like this:

CUP BOTTOM
RIM
GLUE
LEAD PROJECTS SLIGHTLY
SMOOTHED EDGE

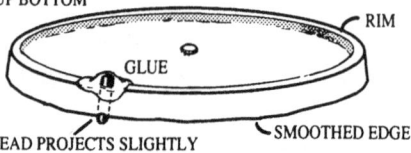

4a. This cardioid is a heart-shaped curve.

4b. A cardioid is traced by a point on the circumference of a circle rolling completely around a fixed circle of equal size.

## Extension

Predict the line that will result from rolling the free disk along a straight-edge. Try it. Are your results a surprise?

## Materials

☐ Two styrofoam cups.
☐ Scissors.
☐ Pencil lead. Thick lead from a wooden pencil stub or thin lead from a mechanical pencil both work.
☐ White glue.
☐ The Cardioid Points supplement.
☐ A ruler or straightedge.

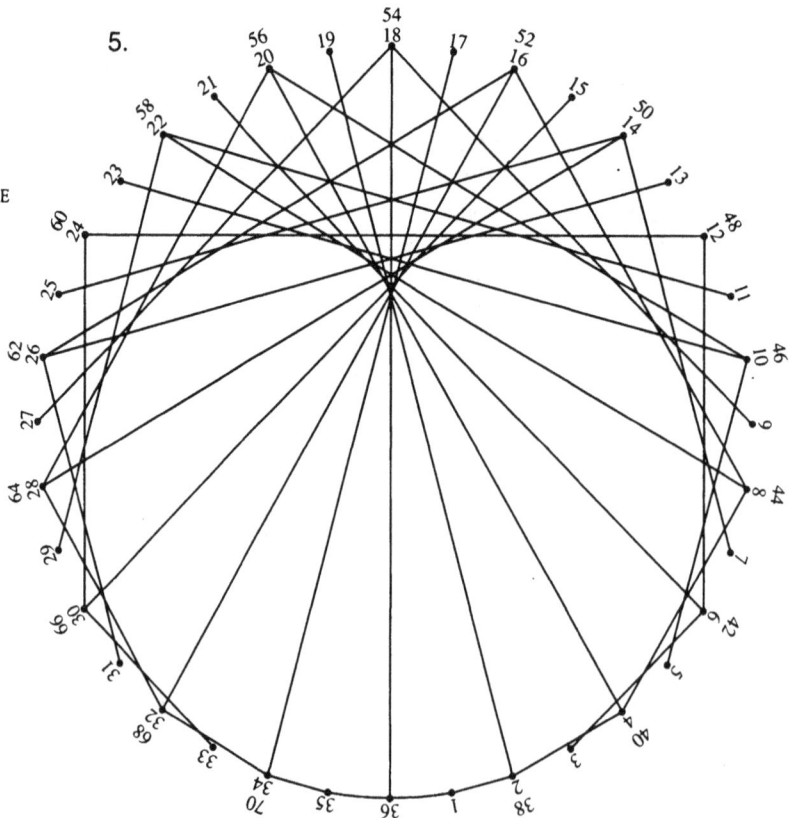

5.

**(TO)** cut and tape the interior angles of polygons, and find their sums. To understand how this sum increases as the number of sides of the polygon increases.

---

## INTERIOR ANGLES ⭘                    Math Lab (   )

1. Join 3 straight lines of any length on a piece of scratch paper to form a large triangle. Cut it out.
   a. Tear off its 3 interior angles, keeping each piece as large as possible.
   b. Join these angles with tiny bits of tape so the sides meet flush and share a common point.
   c. What part of a 360° circle do these interior angles add up to?

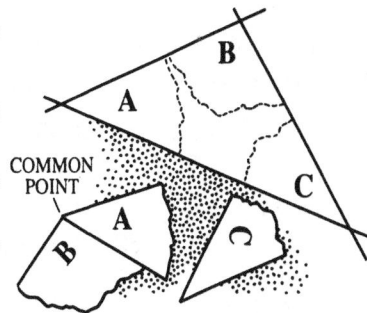

2. Connect 4 straight lines to form a *large* 4-sided polygon. Cut, tear and tape as before. What part of a 360° circle do its interior angles add up to?

3. Repeat with a *large* 5-sided polygon. Hold the overlapping angles up to strong light to discover their sum.

4. What is the sum of the interior angles of this 6-sided polygon? Make a prediction. Explain your reasoning.

5. Test your prediction: draw, cut, tear and tape together a 6-sided polygon.

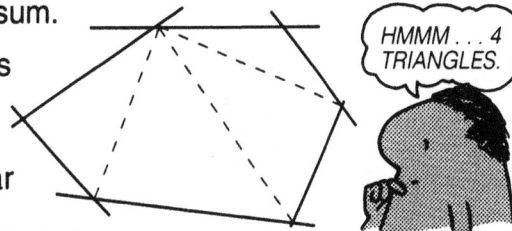

*HMMM . . . 4 TRIANGLES.*

© 1994 by TOPS Learning Systems                                                        13

---

## Answers / Notes

1c. The interior angles of any triangle form a 180° straight line, half of a 360° circle.

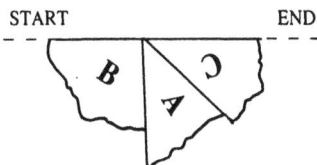

2. The interior angles of any 4-sided polygon add up to a 360° circle.

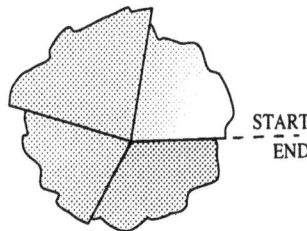

3. The interior angles of any 5-sided polygon add up to 540°, a circle and a half. The leading edge and final edges overlap directly opposite each other, forming a straight line.

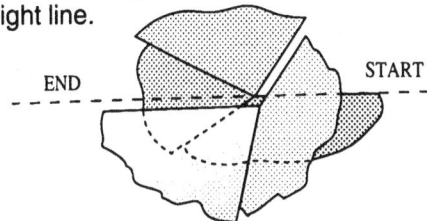

4. The interior angles of this 6-sided polygon should add up to two complete circles (720°). There are 2 good reasons for making this prediction.
   • Generalizing from previous polygons:

   number of sides =   3,   4,    5,    6…
   sum of interior angles = 180°, 360°, 540°, <u>720°</u>…

   • The 6-sided polygon divides into 4 triangles with interior angles that add up to 180° each:

   $$180° \times 4 = \underline{720°}$$

5. The interior angles of a 6-sided polygon did form two complete circles (720°).

*Note: In this activity, we have drawn all polygons with interior angles less than 180°, as if they were inscribed in an ovoid. You might direct younger students to do this so the resulting interior angles will be easy to tear and tape. Students who draw polygons without this constraint may produce highly irregular shapes that are more difficult to join.*

EASY ANGLES

DIFFICULT ANGLE

## Materials
☐ Four sheets of scratch paper per student.
☐ A straight edge, scissors and clear tape.

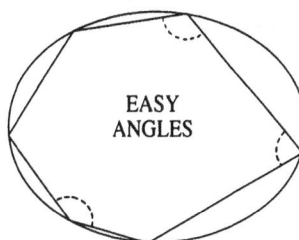

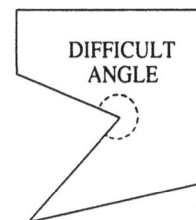

**(TO)** discover number patterns that simplify the addition of long sequences of numbers.

---

### EASY ADDITION ○ Math Lab (   )

1. The first 100 integers (whole numbers) are written below. Find a pattern, then add them all together the easy way.

01 02 03 04 05 06 07 08 09 10 11 12 13 14 15 16 17 18 19 20 21 22 23 24 25
100 99 98 97 96 95 94 93 92 91 90 89 88 87 86 85 84 83 82 81 80 79 78 77 76

26 27 28 29 30 31 32 33 34 35 36 37 38 39 40 41 42 43 44 45 46 47 48 49 50
75 74 73 72 71 70 69 68 67 66 65 64 63 62 61 60 59 58 57 56 55 54 53 52 51

2. What is the sum of the first 1000 integers? Show your work.

3. Add up each horizontal string of 37's.
Why is this easier than it looks?

37 37 37 = ?
37 37 37 37 37 37 = ?
37 37 37 37 37 37 37 37 37 = ?
37 37 37 37 37 37 37 37 37 37 37 37 = ?
37 37 37 37 37 37 37 37 37 37 37 37 37 37 37 = ?
37 37 37 37 37 37 37 37 37 37 37 37 37 37 37 37 37 37 = ?
37 37 37 37 37 37 37 37 37 37 37 37 37 37 37 37 37 37 37 37 37 = ?
37 37 37 37 37 37 37 37 37 37 37 37 37 37 37 37 37 37 37 37 37 37 37 37 = ?
37 37 37 37 37 37 37 37 37 37 37 37 37 37 37 37 37 37 37 37 37 37 37 37 37 37 37 = ?

14

---

### Answers / Notes

1. The first 100 integers make 50 pairs. Each pair adds up to 101. The sum, therefore is:

    101 x 50 = 5000 + 50 = 5050.

2. The first 1000 integers group into 500 pairs totaling 1001 each:

    1000   999   998
    + 1   + 2   + 3   ... = 500 pair.
    ‾‾‾‾   ‾‾‾   ‾‾‾
    1001, 1001, 1001,

    The sum, therefore is 1001 x 500 = 500,000 + 500 = 500,500.

3. 37 + 37 + 37 = 3 x 37 = 111
        6 x 37 = 111 + 111 = 222
        9 x 37 = 111 + 111 + 111 = 333
        ... etc.             = 444
                             = 555
                             = 666
                             = 777
                             = 888
                             = 999

Adding the number 111 to itself forms a number pattern as easy a counting. The numbers don't carry until they exceed 999.

### Materials

☐ None.

**(TO)** list all prime integers under 100 and examine their properties.

---

## PRIME NUMBERS ○ Math Lab ( )

1. Get a <u>Number Sieve.</u>
   a. Copy the definitions for *prime numbers* and *composite numbers.*
   b. Give an example of each.

2. Run the first 100 integers through this sieve so only primes are left behind. To do this…
   a. Cross out all composite numbers that are multiples of the four circled primes.
   b. All numbers that remain in the sieve are prime. List them as a sequence of numbers.
   c. Consecutive odd pairs are called *twin primes.* Circle all twin primes in your list.

| 1 | ⑦ | 13 |
|---|---|---|
| ② | 8 | 14→ |
| ③ | 9 | 15 |
| 4 | 10 | 16→ |
| ⑤ | 11 | 17 |
| 6 | 12 | 18 |

3. Every composite number has a unique set of 2 or more prime factors. What primes multiplied together equal the composite 210?

$$(\text{prime})\,(\text{prime})\ldots = \text{composite}$$

4. Every <u>EVEN number</u> greater than 4 is the <u>sum of two odd primes</u>. How many pairs of odd primes can you find that add up to 90?
   Example: 43 + 47 = 90

5. Every <u>ODD number</u> greater than 7 is the <u>sum of three odd primes</u>. How many trios of odd primes can you find that add up to 89?
   Example: 3 + 3 + 83 = 89

15

---

## Answers / Notes

1a. A prime number is divisible only by itself and the number 1. A composite number is a multiple of smaller primes.

1b. The number 19 is prime because only 1 and 19 divide into it evenly.

The number 18 is a composite because it is a multiple of lower primes (2 x 3 x 3).

2a.

*This method of filtering out primes is known as the "Sieve of Eratosthenes." To capture all the primes in N numbers, only composites of primes equal to or less than √N need to be crossed out.*

Caution students not to overlook 91, a composite of 7.

2b-c. 2, (3, 5), 7, (11, 13), (17, 19), 23, (29, 31), 37, (41, 43), 47, 53, 59, 61, 67, (71, 73), 79, 83, 89, 97.

3. 2 x 3 x 5 x 7 = 210.

4-5. *These assertions were first proposed by the Prussian mathematician Christian Goldbach in 1742. Computer searches have verified that both Goldbach conjectures are true for all integers up to very large numbers, but neither has yet been proven true for all integers. There are a surprisingly large number of solutions for 90 and 89. Your students will not likely find all of them, unless they do a systematic search as below.*

### 90 as the sum of 2 odd primes

| ~~2~~ | 3 | 5 | 7 | 11 | 13 | 17 | 19 | 23 | 29 | 31 | 37 | 41 | 43 | 47 | 53 | 59 | 61 | 67 | 71 | 73 | 79 | 83 | ~~89~~ ~~97~~ |
|---|---|---|---|---|---|---|---|---|---|---|---|---|---|---|---|---|---|---|---|---|---|---|---|
| | | | 83 | 79 | | 73 | 71 | 67 | 61 | 59 | 53 | | 47 | | | | | | | | | | |

— Duplicates above 43. —

### 89 as the sum of three odd primes:

| ~~97~~ | ~~89~~ | 83 | 79 | 73 | 71 | 67 | 61 | 59 | 53 | 47 | 43 | 41 | 37 | 31 | 29 | 23 | 19 | 17 | 13 | 11 | 7 | 5 | 3 | ~~2~~ |
|---|---|---|---|---|---|---|---|---|---|---|---|---|---|---|---|---|---|---|---|---|---|---|---|---|
| | | 3+3 | 3+7 | 13+3 | 13+5 | 19+3 | 23+5 | 23+7 | 31+5 | 37+5 | 43+3 | ~~43+5~~ | ~~47+5~~ | ~~53+5~~ | | | | | | | | | | |
| | | 5+5 | 11+5 | 11+7 | | 17+11 | 17+11 | 19+11 | 29+7 | 31+11 | 41+5 | 41+7 | 29+23 | ~~47+11~~ | | | | | | | | | | |
| | | | | 11+11 | | | 17+13 | 23+13 | 29+17 | 41+11 | 37+11 | ~~41+17~~ | | | | | | | | | | | | |
| | | | | | | | | 19+17 | 23+19 | 23+23 | 31+17 | 29+29 | | | | | | | | | | | | |
| | | | | | | | | | | 29+19 | | | | | | | | | | | | | | |

— Duplicates below 31. —

## Materials

☐ The Number Sieve supplement.

**(TO)** develop a table of logarithms that simplify mathematical computations. To practice four basic log operations.

---

## LOGARITHMS ○ Math Lab ( )

| | n | log n | |
|---|---|---|---|
| MULTIPLY | 4 | 2 | ADD |
| | × 8 | +3 | |
| PRODUCT = | 32 | 5 | = SUM |

1. Copy this table on notebook paper, using the margin line as a divider. Follow the pattern until you end with 1,048,576 and 20.

2. Notice how the *product* of numbers on the left match the *sum* of their logs on the right. Show that this pattern holds for at least 3 other problems.

3. Use your log table to solve the boxed problems like each example. Check each answer with a calculator.

*You may use a calculator.*

| n | log n |
|---|---|
| 1 | 0 |
| 2 | 1 |
| 4 | 2 |
| 8 | 3 |
| 16 | 4 |
| 32 | 5 |
| | 6 |
| | 18 |
| 524,288 | 19 |
| 1,048,576 | 20 |

**a. Multiply: | Add:**
4⊗8 = 32 | 2⊕3 = 5

$$32 \times 64 = ?$$
$$128 \times 2048 = ?$$
$$4 \times 16 \times 512 = ?$$

**b. Divide: | Subtract:**
16⊘8 = 2 | 4⊖3 = 1

$$1,024 \div 128 = ?$$
$$131,072 \div 65,536 = ?$$
$$8,192 \div 32 \div 64 = ?$$

**c. Power: | Multiply:**
4③ = 64 | 2⊗3 = 6

$$32^2 = ?$$
$$16^3 = ?$$
$$(8^2)^3 = ?$$

**d. Index: | Divide:**
②√64 = 8 | 6⊘2 = 3

$$^2\sqrt{16,384} = ?$$
$$^3\sqrt{262,144} = ?$$
$$^4\sqrt{1,048,576} = ?$$

© 1994 by TOPS Learning Systems

16

---

## Introduction

$3^2 = 3 \times 3$. Thus, $3^2 = 9$.  
$3^3 = 3 \times 3 \times 3$. Thus $3^3 = 27$.  
$3^4 = 3 \times 3 \times 3 \times 3$. Thus $3^4 = 81$.

$8 \times 8 = 64$. Thus $^2\sqrt{64} = 8$.  
$4 \times 4 \times 4 = 64$. Thus $^3\sqrt{64} = 4$.  
$2 \times 2 \times 2 \times 2 \times 2 \times 2 = 64$. Thus $^6\sqrt{64} = 2$.

## Answers / Notes

1.

| n | log n |
|---|---|
| 1 | 0 |
| 2 | 1 |
| 4 | 2 |
| 8 | 3 |
| 16 | 4 |
| 32 | 5 |
| 64 | 6 |
| 128 | 7 |
| 256 | 8 |
| 512 | 9 |
| 1,024 | 10 |
| 2,048 | 11 |
| 4,096 | 12 |
| 8,192 | 13 |
| 16,384 | 14 |
| 32,768 | 15 |
| 65,536 | 16 |
| 131,072 | 17 |
| 262,144 | 18 |
| 524,288 | 19 |
| 1,048,576 | 20 |

2. *Students should supply at least 3 examples. Products to the left always match sums to the right within the capacity of the table.*

3a.
$5 + 6 = 11$ | 2,048 [Check: 32 x 64 = 2,048]
$7 + 11 = 18$ | 262,144 [Check: 128 x 2,048 = 262,144]
$2 + 4 + 9 = 15$ | 32,768 [Check: 4 x 16 x 512 = 32,768]

3b.
$10 - 7 = 3$ | 8 [Check: 1,024 ÷ 128 = 8]
$17 - 16 = 1$ | 2 [Check: 131,072 ÷ 65,536 = 2]
$13 - 5 - 6 = 2$ | 4 [Check: 8,192 ÷ 32 ÷ 64 = 4]

3c.
$5 \times 2 = 10$ | 1,024 [Check: $32^2 = 1,024$]
$4 \times 3 = 12$ | 4,096 [Check: $16^3 = 4,096$]
$3 \times 2 \times 3 = 18$ | 262,144 [Check: $(8^2)^3 = 262,144$]

3d.
$14 \div 2 = 7$ | 128 [Check: $^2\sqrt{16,384} = 128$]
$18 \div 3 = 6$ | 64 [Check: $^3\sqrt{262,144} = 64$]
$20 \div 4 = 5$ | 32 [Check: $^4\sqrt{1,048,576} = 32$]

## Materials

☐ Notebook paper.  
☐ A hand calculator.

**(TO)** discover number patterns in a slide and jump game. To generalize an nth term for each pattern.

---

**SLIDE AND JUMP**　　　　　　　○　　　　　　　　**Math Lab (　)**

1. Stick masking tape across your table as long as this paper. Draw a 3-box "game board" in the middle with a penny in the left box, a paper clip in the right.

2. How many single slides plus single jumps does it take to move the markers to opposite sides of the board? (Pennies move right *only*; paper clips move left *only*.) Copy this table and complete the first entry.

3. Expand to a 5-square game board with 2 pennies (and 2 paper clips). Play, then fill in the next entry. Apply the rule of thumb to keep from getting trapped.　➡

4. Complete your table through 6 pennies.
   a. Describe the sequence in each row.
   b. Complete a final entry for n pennies.

| pennies | 1 | 2 | 3 | 4 | 5 | 6 | n |
|---------|---|---|---|---|---|---|---|
| squares | 3 | 5 |   |   |   |   |   |
| moves   |   |   |   |   |   |   |   |

**Rule of Thumb:**
Never place markers of the same kind side by side (except in starting and ending positions).

© 1994 by TOPS Learning Systems　　　　17

---

## Answers / Notes

2. Three moves.
   ┌ penny slides right
   │ paper clip jumps left
   │ penny slides right
   └
   *Could also start by sliding the paper clip left.*

3. Eight moves.
   penny slides right
   paper clip jumps left
   paper clip slides left
   penny jumps right
   penny jumps right
   paper clip slides left
   paper clip jumps left
   penny slides right

*There is a strong tendency to continue trading moves between the paper clip and penny. (This worked the first time.) If you do slide the penny right in this step, however, then a paper clip must jump left in the following step. Two paper clips end up side by side, violating the rule of thumb and resulting in a dead-end game.*

4.

| pennies | 1 | 2 | 3 | 4 | 5 | 6 | n |
|---------|---|---|---|----|----|----|--------|
| squares | 3 | 5 | 7 | 9 | 11 | 13 | $2n + 1$ |
| moves   | 3 | 8 | 15 | 24 | 35 | 48 | $n^2 + 2n$ |

top row: each new term is one more than the previous term (like counting).

middle row: each new term is 2 more than the previous term (like counting with odd numbers).

bottom row: The difference between succeeding pairs of terms increases by 2.

*The general term for the "moves" sequence is difficult to guess. Try leading your class to the answer with a table like this:*

1, 2, 3, 4, 5,...　　n
1, 4, 9, 16, 25,...　　$n^2$
4, 9, 16, 25, 36,...　　$(n+1)^2$
3, 8, 15, 24, 35,...　　$(n+1)^2 - 1$
　　　　　$= (n^2 + 2n + 1) - 1$
　　　　　$= n^2 + 2n$

## Materials

☐ Masking tape.
☐ Six pennies and six paper clips.

## Extension

Recording consecutive penny moves and paper clip moves as separate totals within each game forms this interesting triangle. What internal patterns do you see?

1+1+1
1+2+2+2+1
1+2+3+3+3+2+1
1+2+3+4+4+4+3+2+1
1+2+3+4+5+5+5+4+3+2+1
1+2+3+4+5+6+6+6+5+4+3+2+1

**(TO)** be surprised by the unusual topological properties of a Mobius strip.

---

## MOBIUS TWIST ○ Math Lab ( )

1. Cut about 1 meter of adding machine tape with neat, square ends…
Hold the ends together in a closed loop. Turn one end over to make a half twist, and tape the ends together all the way across to make a *Mobius strip*.

1/2 TWIST   TAPE

2. Write numbers along the middle of the strip (1, 2, 3…) all the way around, spaced 1 hand width apart. Argue that this Mobius strip has only 1 side!

1   2   3   4

3. Poke a pinhole near the edge *above* each number. Argue that a Mobius strip has only 1 edge!

9  10  11  12  13  14
PINHOLES

4. Cut between the numbers and the pinholes all the way around. What is the surprise? Is either piece still a Mobius strip? Explain.

15   16   17   18

5. Carefully tear the Mobius strip through the center of all its numbers. What happens? Is the torn piece still a Mobius strip? Explain.

© 1994 by TOPS Learning Systems                                    18

### Answers / Notes

2. Numbering around the strip, you return to the number you started with. The strip was never lifted from the table, yet no second side remains to be numbered!

3. Poking pinholes near the edge of the strip above each number, you return to the top of the number you started with. The first edge was never abandoned, yet no second edge remains to be poked!

4. The Mobius strip divides into 2 twisted interlocking loops, one twice as long as the other.

The longer (2 meter) loop is no longer a Mobius strip. It now has both a pinhole edge and a cut edge. Moreover, you can number or draw on it all the way around and a second side remains unmarked. *(Students will likely disagree about the properties of this strip; encourage constructive class debate. Settle the question with experiments that students suggest.)*

The shorter (1 meter) loop is still a Mobius strip. Its single numbered side remains intact. Its new edge was made with one continuous cut of the scissors. (Or, if you poke new pinholes near the edge, above each number, no second edge remains to be poked.)

5. The Mobius strip divides into 1 loop that is twice as long as the original. This loop is no longer a Mobius strip because it now has two edges: one cut, the other torn. It also has two sides: draw a line on one side without lifting your pencil and a second side remains unmarked.

### Extensions

1. Lightly tape the interlocking loops back into your original Mobius strip. (This is extremely difficult. A good strategy is to tear a new Mobius strip, as before, paying close attention to how it falls apart.)

2. Cut a slit into both ends of a 50 cm strip of adding machine tape. Tape the ends together like this. Does this paper have Mobius properties? (Yes. It has 1 edge and 1 side.)

SLIT BOTH ENDS
OVERLAP & TAPE

3. Make a strip as in step 1 but with a *full* twist before taping. Test its properties.

### Materials

☐ Adding machine tape.
☐ A meter stick.
☐ Scissors.
☐ A straight pin.

**(TO)** interpret bar codes used by the U.S. postal service that automate mail sorting.

---

## POSTAL BAR CODES     ◯     Math Lab (   )

1. Draw vertical lines on lined paper to form a grid 10 rows deep and 5 squares wide.

   a. Fill in the top row with 2 tall and 3 short bars like these.

   b. Fill in the remaining 9 rows, always with 2 tall and 3 short bars, but in *different* sequences.

2. Write these numbers above the five columns. Add up the value of the *long* bars on each line (short bars don't count). Write totals to the right.

3. Get a <u>Bar</u> <u>Code</u>. All codes and their numbers along the bottom match the totals in your table, except one. Which has been assigned a new value?

4. The correction character (last box) is chosen so all the numbers add up to a multiple of 10. Show that this is true.

5. Write correct bar codes for these addresses:

| TOPS Learning Systems 10970 S. Mulino Rd Canby OR 97013-8799 | Jo Smith PO Box 345 LA CA 90027-3618 | Your Name Address City State Zip |
|---|---|---|

6. Decode bars cut from different envelopes. Label each answer with the circled letter on the envelope.

© 1994 by TOPS Learning Systems                                    19

---

## Answers / Notes

*1-2. These rows of long and short bars may be arranged in any order.*

3. The number 11 (2 long bars followed by 3 short) has been reassigned a value of zero.

Numbers add to 60 with this 0.

4.
```
   1
   2
   3
   4
   5
   6
   7
   8
   9
   0
   1
 + 4
 ────
  50 (a multiple of 10)
```

5. TOPS = |||||||||||||||||||||||||||||||||||||||||||||||||||||||

   Smith = |||||||||||||||||||||||||||||||||||||||||||||||||||||||

   Student address = various answers

Numbers add to 50 with this 5.

6. Various answers.

## Materials

☐ Lined notebook paper.

☐ The Bar Code supplement.

☐ Several envelopes per student or lab group, imprinted with full 62-line bar codes (12 digits plus frame bars). There are shorter versions without the +4 zip and/or 2 digit delivery point. Label each address and its bar code with the same circled letter, then cut them apart. When students come to you for a checkpoint, confirm that their decoded bar numbers match the address numbers on the envelope with the same circled letter.

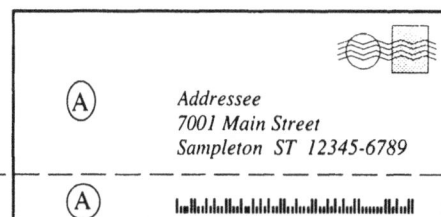

Ⓐ Addressee
7001 Main Street
Sampleton ST 12345-6789

Ⓐ |||||||||||||||||||||||||||||||||||||||||||||

**(TO)** cut out 12 unique 5-tile puzzle pieces and use them to construct rectangles of various dimensions.

## JIGSAW PUZZLES

○                                Math Lab (   )

1. Fold the <u>Centimeter</u> <u>Grid</u> in half at the arrows, so all lines match when you hold the paper up to strong light. Stick together with a thin, even layer of glue.

2. Only five of these shapes are unique.

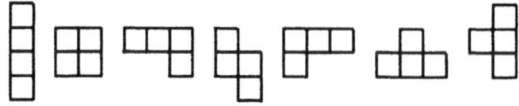

Find 2 shapes that are duplicated…
a. by rotation.     b. by mirror image.

3. Cut 12 unique shapes out of your glued squares that have FIVE tiles each. Record the shape of each piece on a new tile sheet.

4. How many different rectangles can you build from these 12 shapes? Record each solution on your tile sheet. (You may turn pieces over.)

© 1994 by TOPS Learning Systems                                20

## Answers / Notes

2a.

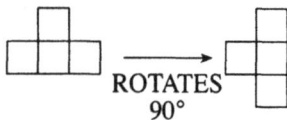

ROTATES 90°

2b.

MIRROR IMAGE

3.

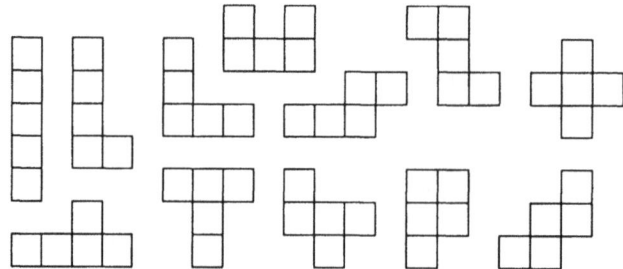

4. *The smallest rectangle students can "build" from these tiles is a trivial 1x5. The larger rectangles become progressively harder to build. Using all the pieces is a major challenge.*

*Here are some of many possible solutions. If you and your students solve 5x12, 4x15 or 3x20 rectangles, please send us your solutions.*

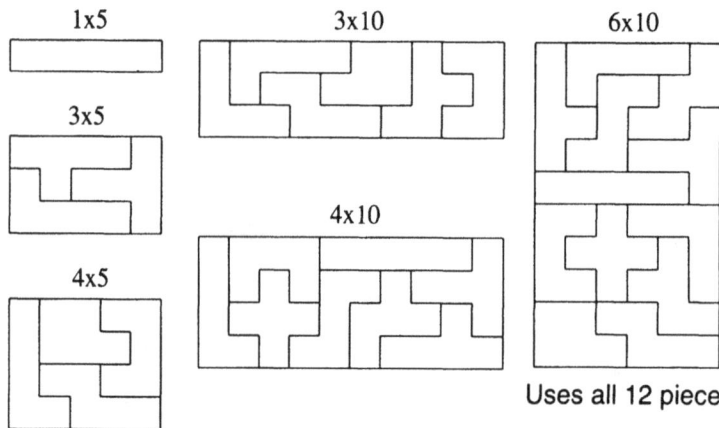

1x5     3x10     6x10

3x5

4x10

4x5

Uses all 12 pieces.

## Materials

☐  Two Centimeter Tiles sheets.
☐  Glue.
☐  Scissors.

# REPRODUCIBLE
# STUDENT
# TASK CARDS

# Task Cards Options

**Here are 3 management options to consider before you photocopy:**

**1. Consumable Worksheets:** Copy 1 complete set of task card pages. Cut out each card and fix it to a separate sheet of boldly lined paper. Duplicate a class set of each worksheet master you have made, 1 per student. Direct students to follow the task card instructions at the top of each page, then respond to questions in the lined space underneath.

**2. Nonconsumable Reference Booklets:** Copy and collate the 2-up task card pages in sequence. Make perhaps half as many sets as the students who will use them. Staple each set in the upper left corner, both front and back to prevent the outside pages from working loose. Tell students that these task card booklets are for reference only. They should use them as they would any textbook, responding to questions on their own papers, returning them unmarked and in good shape at the end of the module.

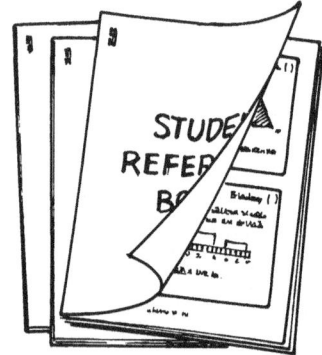

**3. Nonconsumable Task Cards:** Copy several sets of task card pages. Laminate them, if you wish, for extra durability, then cut out each card to display in your room. You might pin cards to bulletin boards; or punch out the holes and hang them from wall hooks (you can fashion hooks from paper clips and tape these to the wall); or fix cards to cereal boxes with paper fasteners, 4 to a box; or keep cards on designated reference tables. The important thing is to provide enough task card reference points about your classroom to avoid a jam of too many students at any one location. Two or 3 task card sets should accommodate everyone, since different students will use different cards at different times.

## PENNY PLAY ⭕        Math Lab (  )

1. These penny patterns form number sequences. Write the first 10 terms of each sequence as a string of numbers.

a.
1, 3, 6,

(The differences between neighboring terms also forms a number pattern.)

b.
1, 4, 9,

(These numbers are called *perfect squares*.)

c.
1, 7, 19,

(Build the next term like this: 1 "heads" surrounded by 6 tails, surrounded by 12 heads, surrounded by more tails...)

2. Use the sequences you have just written to help you find the 10th term in each of these sequences:

    a. 1, 1+2, 1+2+3, ...
    b. 1, 1+3, 1+3+5, ...
    c. 1, 1+7, 1+7+19, ...

3. Here is a penny puzzle: This triangle points up. Move just 3 pennies to make it point down.

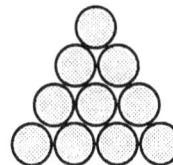

© 1994 by TOPS Learning Systems     1

---

## GENERAL TERM ⭕        Math Lab (  )

For each sequence below...

1. Draw the 4th term, showing 4 shaded squares.

2. Write a general term (**n**th term) for **n** shaded squares.

3. Use this general term to find the 5th term.

4. Write the first 5 terms as numbers. Check that the differences between terms also forms a pattern.

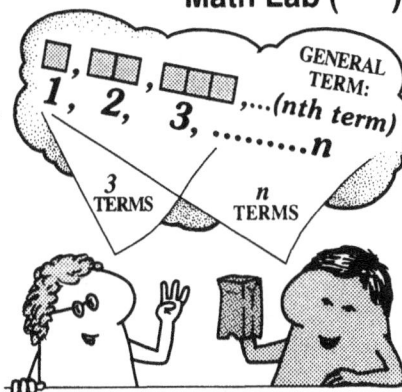

a. , ...

b. , ...

c. , ...

d. , ...

e. , ...

f. , ...

© 1994 by TOPS Learning Systems     2

## THE FIBONACCI SEQUENCE  ○                  Math Lab (    )

1. The Italian mathematician Leonardo Fibonacci first
investigated this sequence of numbers in 1683.

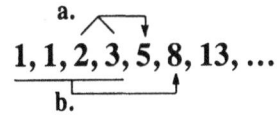

a.
$$1, 1, 2, 3, 5, 8, 13, \ldots$$
b.

   a. How are the previous 2 terms always related to
the next term? Use your rule to extend this sequence to 15 terms.
   b. How are the *sum of the first n terms* in this sequence related to the
number that follows 2 terms later?

2. An alien life form, a juvenile **b**<u>lob</u>, reaches Earth in a space pod.
In 1 day it grows to an adult **B**lob (as big as a pea). In 1 more day (and
every day thereafter) this adult Blob gives birth to another juvenile
blob, each of which matures and reproduces like its parent.

   a. Track the blob population for *1 week*
on lined paper, like this.
   b. What is the total blob population
after the 30th day?
   c. Each blob consumes 1 gram of
biomass per day. How much does this
alien life form consume after 30 days?
(1,000,000 grams = 1 metric ton)

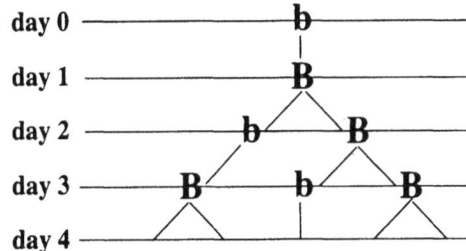

day 0 ——————— **b**
day 1 ——————— **B**
day 2 ——— **b** —— **B**
day 3 — **B** —— **b** —— **B**
day 4 ———————

3

---

## SEASHELL SPIRAL                    ○                  Math Lab (    )

1. <u>Draw</u> a 14.4 cm square on paper with square corners. Use an index card to make right angles.
   a. <u>Inscribe</u> a quarter circle from corner to corner with a compass.
   b. <u>Cut</u> out the square.

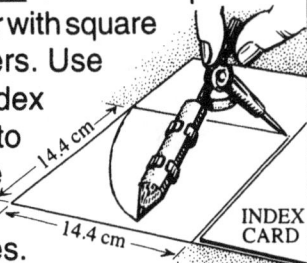

2. <u>Draw</u>, <u>inscribe</u>, and <u>cut</u> smaller squares of 8.9 cm, 5.5 cm, 3.4 cm, 2.1 cm, and 1.3 cm from your paper. Tape them together so the curved lines spiral inward.

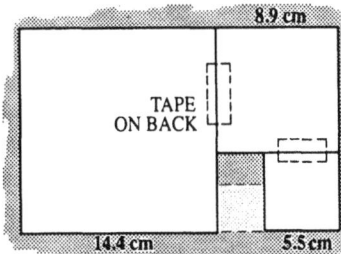

3. Tape paper *behind* the hole that remains. Draw and inscribe smaller squares so the line spirals to a point.

8.9 cm
TAPE ON BACK
14.4 cm        5.5 cm

PATCH HOLE FROM BACK

4. These squares define the shape of a chambered nautilus. What determines
the size of each square?

5. Where else can you find Fibonacci numbers?

COSMOS          GERANIUM LEAF      PIANO KEYS
PERIWINKLE

4

## THE GOLDEN MEAN  ○  Math Lab (   )

1. Measure the length (**L**) and width (**W**) of all 8 <u>Golden</u> <u>Rectangles</u> to the nearest whole mm. Use a calculator (to 3 decimal places) to complete this table.

| L(mm) | W(mm) | L+W | $\frac{W}{L}$ | $\frac{L}{L+W}$ |
|-------|-------|-----|-----|-----|
| 233 | 144 | 377 | .618 | .618 |

↓ (List largest to smallest.) ↓

  a. Do you recognize the numbers in the first 3 columns of your table?
  b. The last 2 columns contain ratios that equal the *golden mean*. Write a definition of the golden mean.
  c. If you were to cut a square off a golden rectangle, what remains?

2. Carefully cut around the outer rectangle. Fold it in half so all edges match.
  a. Snip into the folded borders of all 7 rectangles.
  b. Unfold, then finish cutting out all 7 "picture frames."

SNIP TO START

PASTE DOWN A DESIGN YOU LIKE!

3. The golden mean has been used in art and architecture since ancient times. Arrange your "golden" frames on dark construction paper to create your own eye-catching design.

5

## CIRCLE SEQUENCE  ○  Math Lab (   )

1. All points on these circles are joined to all other points by straight lines. This makes a sequence of increasing points and regions.

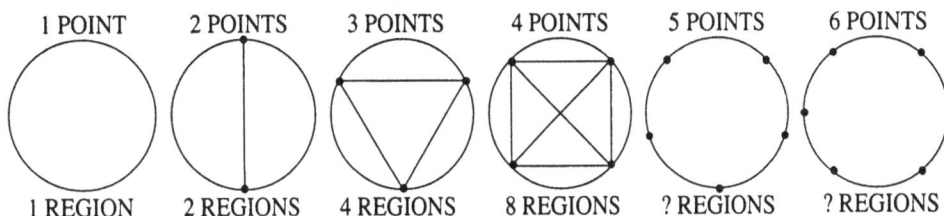

| 1 POINT | 2 POINTS | 3 POINTS | 4 POINTS | 5 POINTS | 6 POINTS |
|---------|----------|----------|----------|----------|----------|
| 1 REGION | 2 REGIONS | 4 REGIONS | 8 REGIONS | ? REGIONS | ? REGIONS |

  a. Predict the number of regions in the 5th and 6th terms.
  b. On what pattern did you base your generalization?

...4,5,6...

2. Use a compass to draw large copies of the circles with 5 and 6 points equally spaced around the outside. Draw straight lines joining every point with *every* other point in the circle.

3. Count the regions in each circle.
  a. Was your <u>prediction</u> correct? Explain.
  b. Was your <u>generalization</u> reasonable? Explain.
  c. What does this activity tell you about predictions and generalizations?

6

## POOL TABLE INDUCTION ○ Math Lab ( )

1. Copy these 6 "pool tables" on a <u>Centimeter</u> <u>Grid</u>.

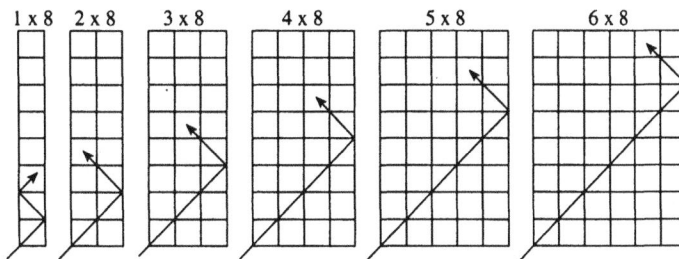

1 x 8  2 x 8   3 x 8    4 x 8     5 x 8      6 x 8

*EVERY BOUNCE FORMS A RIGHT ANGLE.*

2. A cue ball is hit from the lower left corner of each table and bounces as shown. Draw the ball's path across each table until it finally drops into a corner pocket. (Assume there are no side pockets.)

   a. In which pocket does the cue ball always drop?
   b. If this trend continues, where should the cue ball drop on a 7x8 and 8x8 pool table? Make a generalization.
   c. Is this generalization certain?

3. Test your prediction.
   a. Draw 7x8 and 8x8 pool tables on graph paper and "shoot" the cue ball. What happens?
   b. What have you learned about making generalizations?

7

---

## TO THE LIMIT ○ Math Lab ( )

4  3  2  1

1. Cut out the <u>Number</u> <u>Square</u>. There are 4 pairs of points in section A labeled 1, 2, 3 and 4. Join each pair with a straight line.

2. Join pairs of points in sections B, C and D in a similar manner with a sharpened pencil.

3. Examine the perimeter of the central shape within each section A, B, C and D. Do you see…

   • straight lines crossing at angles? (or)
   • a smooth curved line without angles?

*HOW SMOOTH IS THIS PERIMETER IN EACH QUADRANT?*

4. Consider this sequence of regular polygons.

△ , □ , ⬠ , ⬡ , ⬡ , ⬡ , ⬡ , ⬡ , …

a. As the number of sides grow without limit, what shape does the polygon eventually assume?

b. How many straight sides are required to form a perfect circle? How long is each side?

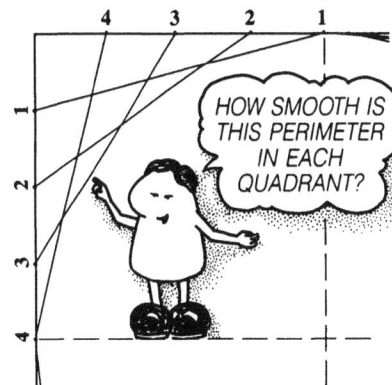

8

## CIRCLE  ○

1. Cut corrugated cardboard to notebook paper size.

2. Cut a piece of thread equal to the width of notebook paper. Tie it in a loop.

3. Center the paper on the cardboard and stick a pin in the middle. Put the loop over the pin.

4. Tighten the loop with your pencil point, and draw a circle around the pin. Mark the center "c" and points $p_1$, $p_2$, $p_3$, $p_4$ like these:

5. Copy this definition: A *circle* is the set of all points in a plane that are the same distance from a fixed point (the center).

   a. What plane do these 5 points share?

   b. Which points "p" are part of the circle? Why?

6. Cut out your circle. Fold over the edge to touch the center, then crease sharply with your fingernail. Do this repeatedly until you define a new shape. What is it?

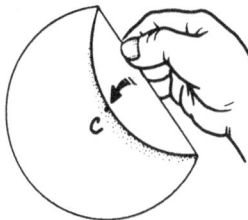

7. Turn the circle over so the creases form raised ridges. Accent the ridges with the *side* of a dark crayon.

© 1994 by TOPS Learning Systems     9

---

## ELLIPSE  ○

1. Set notebook paper on cardboard as before. Lay your thread loop in a circle centered on the paper.

2. Stick 2 pins just *inside* the loop at opposite sides of the circle.

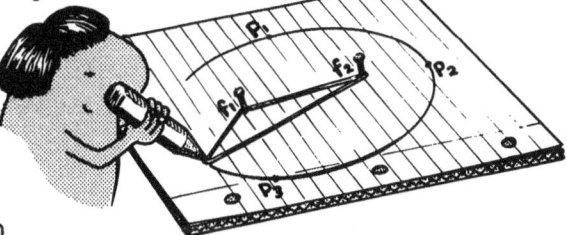

3. Tighten the loop with your pencil point and draw around both pins. Mark the foci $f_1$ and $f_2$ at each pin, and points $p_1$, $p_2$ and $p_3$ like these:

4. Copy this definition: An *ellipse* is the set of all points in a plane such that the sum of the distance of each point from two fixed points (called the foci) is the same. Measure to each point "p" to show this is true.

5. Cut out your ellipse. Fold the edge in to touch focus $f_1$ (not $f_2$), then crease the fold. Repeat over and over until you define a new shape.

6. Darken the ridges on the back with crayon as before. Describe the shape that emerges.

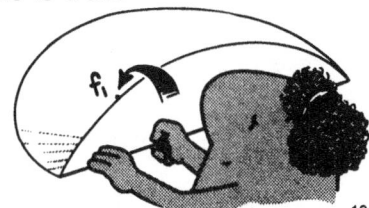

© 1994 by TOPS Learning Systems     10

## PARABOLA     O     Math Lab (    )

1. Carefully trim notebook paper along its margin line. Center a pinhole 4 line widths inside this margin. Label it "f" for focus.

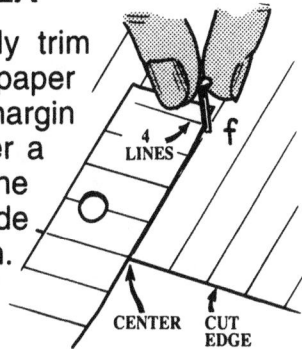

4 LINES    f

CENTER    CUT EDGE

2. Fold over this margin edge to touch the focus point, then crease the fold. Repeat all along the margin until the creases define a *parabola*. Shade the ridges on the back with crayon as before.

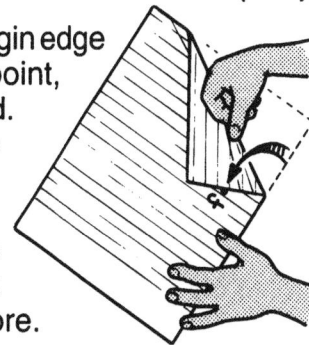

3. Tie thread around a pin. Leave one end as long as your paper. Trim off the other end.

4. Poke the pin into your curved *parabola* at different points.

   a. At each point, use thread to compare distances from pinhead to focus; from pinhead to margin.

   b. Finish this definition: A *parabola* is the set of all points in a plane such that...

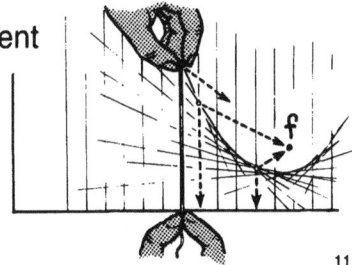

11

---

## CARDIOID     O     Math Lab (    )

CUT SIDES AWAY     TURN OVER
SMOOTH EDGES

1. Cut away the sides of two styrofoam cups. Smooth the upper surfaces of the remaining disks. Turn these so the rims face up.

2. Poke a short piece of pencil lead through the rim of one disk so it sticks out about 1 mm beyond the flat side. Glue it in place on the rim.

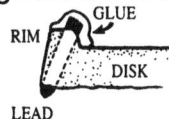

RIM    GLUE    DISK    LEAD

3. Lightly glue the flat side of the other disk to the center of some paper.

4. Roll the free disk all the way around the fixed disk so the pencil lead leaves a smooth line. Roll the disks against each other—don't let them slide.

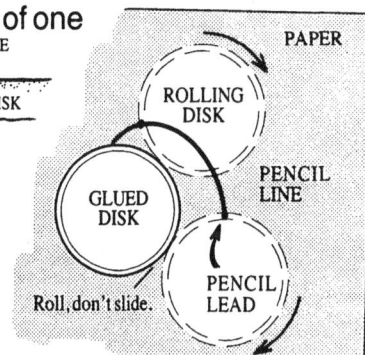

   a. Describe the *cardioid* that you have drawn.
   b. Write a definition for this term.

PAPER    ROLLING DISK    PENCIL LINE    GLUED DISK    PENCIL LEAD    Roll, don't slide.

5. Get a set of <u>Cardioid Points</u>. Connect them with straight lines so each point is joined to a number that is twice its value: 1 to 2, 2 to 4, 3 to 6, 4 to 8,....

12

## INTERIOR ANGLES  ◯

1. Join 3 straight lines of any length on a piece of scratch paper to form a large triangle. Cut it out.
   a. Tear off its 3 interior angles, keeping each piece as large as possible.
   b. Join these angles with tiny bits of tape so the sides meet flush and share a common point.
   c. What part of a 360° circle do these interior angles add up to?

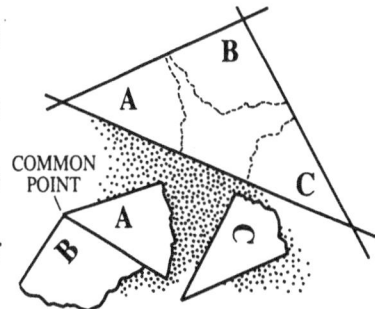

COMMON POINT

2. Connect 4 straight lines to form a *large* 4-sided polygon. Cut, tear and tape as before. What part of a 360° circle do its interior angles add up to?

3. Repeat with a *large* 5-sided polygon. Hold the overlapping angles up to strong light to discover their sum.

4. What is the sum of the interior angles of this 6-sided polygon? Make a prediction. Explain your reasoning.

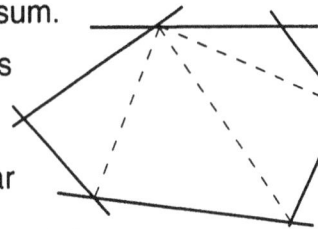

5. Test your prediction: draw, cut, tear and tape together a 6-sided polygon.

HMMM . . . 4 TRIANGLES.

© 1994 by TOPS Learning Systems     13

---

## EASY ADDITION  ◯

1. The first 100 integers (whole numbers) are written below. Find a pattern, then add them all together the easy way.

```
01 02 03 04 05 06 07 08 09 10 11 12 13 14 15 16 17 18 19 20 21 22 23 24 25
100 99 98 97 96 95 94 93 92 91 90 89 88 87 86 85 84 83 82 81 80 79 78 77 76

26 27 28 29 30 31 32 33 34 35 36 37 38 39 40 41 42 43 44 45 46 47 48 49 50
75 74 73 72 71 70 69 68 67 66 65 64 63 62 61 60 59 58 57 56 55 54 53 52 51
```

2. What is the sum of the first 1000 integers? Show your work.

3. Add up each horizontal string of 37's. Why is this easier than it looks?

```
37 37 37 = ?
37 37 37 37 37 37 = ?
37 37 37 37 37 37 37 37 37 = ?
37 37 37 37 37 37 37 37 37 37 37 37 = ?
37 37 37 37 37 37 37 37 37 37 37 37 37 37 37 = ?
37 37 37 37 37 37 37 37 37 37 37 37 37 37 37 37 37 37 = ?
37 37 37 37 37 37 37 37 37 37 37 37 37 37 37 37 37 37 37 37 37 = ?
37 37 37 37 37 37 37 37 37 37 37 37 37 37 37 37 37 37 37 37 37 37 37 37 = ?
37 37 37 37 37 37 37 37 37 37 37 37 37 37 37 37 37 37 37 37 37 37 37 37 37 37 37 = ?
```

© 1994 by TOPS Learning Systems     14

## PRIME NUMBERS $\qquad$ O $\qquad$ Math Lab ( )

**1. Get a Number Sieve.**

   a. Copy the definitions for *prime numbers* and *composite numbers.*

   b. Give an example of each.

**2. Run the first 100 integers through this sieve so only primes are left behind. To do this…**

   a. Cross out all composite numbers that are multiples of the four circled primes.

   b. All numbers that remain in the sieve are prime. List them as a sequence of numbers.

   c. Consecutive odd pairs are called *twin primes.* Circle all twin primes in your list.

| | | |
|---|---|---|
| 1 | (7) | 13 |
| (2) | 8 | 14 |
| (3) | 9 | 15 |
| 4 | 10 | 16 |
| (5) | 11 | 17 |
| 6 | 12 | 18 |

**3.** Every composite number has a unique set of 2 or more prime factors. What primes multiplied together equal the composite 210?

$$(prime)\ (prime)\ … \ = composite$$

**4.** Every <u>EVEN number</u> greater than 4 is the <u>sum of two odd primes</u>. How many pairs of odd primes can you find that add up to 90?

   Example: 43 + 47 = 90

**5.** Every <u>ODD number</u> greater than 7 is the <u>sum of three odd primes</u>. How many trios of odd primes can you find that add up to 89?

   Example: 3 + 3 + 83 = 89

15

---

## LOGARITHMS $\qquad$ O $\qquad$ Math Lab ( )

**1.** Copy this table on notebook paper, using the margin line as a divider. Follow the pattern until you end with 1,048,576 and 20.

*You may use a calculator.*

| $n$ | $\log n$ |
|---|---|
| 1 | 0 |
| 2 | 1 |
| 4 | 2 |
| 8 | 3 |
| 16 | 4 |
| 32 | 5 |
| | 6 |
| | 18 |
| 524,288 | 19 |
| 1,048,576 | 20 |

**2.** Notice how the *product* of numbers on the left match the *sum* of their logs on the right. Show that this pattern holds for at least 3 other problems.

| | $n$ | $\log n$ | |
|---|---|---|---|
| MULTIPLY | 4 | 2 | ADD |
| | × 8 | + 3 | |
| PRODUCT = | 32 | 5 | = SUM |

**3.** Use your log table to solve the boxed problems like each example. Check each answer with a calculator.

**a. Multiply: | Add:**

$4 \otimes 8 = 32$    $2 \oplus 3 = 5$

| 32 x 64 = ? |
| 128 x 2048 = ? |
| 4 x 16 x 512 = ? |

**b. Divide: | Subtract:**

$16 \oslash 8 = 2$    $4 \ominus 3 = 1$

| 1,024 + 128 = ? |
| 131,072 ÷ 65,536 = ? |
| 8,192 ÷ 32 ÷ 64 = ? |

**c. Power: | Multiply:**

$4^{\circledS} = 64$    $2 \otimes 3 = 6$

| $32^2$ = ? |
| $16^3$ = ? |
| $(8^2)^3$ = ? |

**d. Index: | Divide:**

$\sqrt[\circledS]{64} = 8$    $6 \oslash 2 = 3$

| $\sqrt[2]{16,384}$ = ? |
| $\sqrt[3]{262,144}$ = ? |
| $\sqrt[4]{1,048,576}$ = ? |

16

## SLIDE AND JUMP  ○  Math Lab (  )

1. Stick masking tape across your table as long as this paper. Draw a 3-box "game board" in the middle with a penny in the left box, a paper clip in the right.

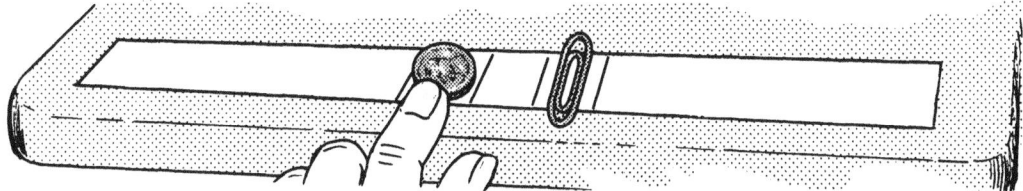

2. How many single slides plus single jumps does it take to move the markers to opposite sides of the board? (Pennies move right *only*; paper clips move left *only*.) Copy this table and complete the first entry.

3. Expand to a 5-square game board with 2 pennies (and 2 paper clips). Play, then fill in the next entry. Apply the rule of thumb to keep from getting trapped. ➡

| pennies | 1 | 2 | 3 | 4 | 5 | 6 | n |
|---------|---|---|---|---|---|---|---|
| squares | 3 | 5 |   |   |   |   |   |
| moves   |   |   |   |   |   |   |   |

4. Complete your table through 6 pennies.
   a. Describe the sequence in each row.
   b. Complete a final entry for n pennies.

**Rule of Thumb:**
Never place markers of the same kind side by side (except in starting and ending positions).

© 1994 by TOPS Learning Systems                    17

---

## MOBIUS TWIST  ○  Math Lab (  )

1. Cut about 1 meter of adding machine tape with neat, square ends…
Hold the ends together in a closed loop. Turn one end over to make a half twist, and tape the ends together all the way across to make a *Mobius strip*.

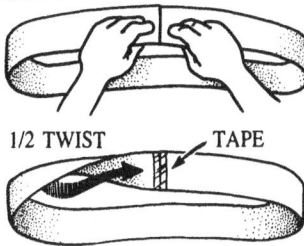

1/2 TWIST      TAPE

2. Write numbers along the middle of the strip (1, 2, 3…) all the way around, spaced 1 hand width apart. Argue that this Mobius strip has only 1 side!

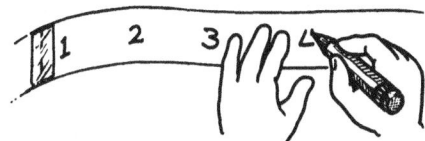

3. Poke a pinhole near the edge *above* each number. Argue that a Mobius strip has only 1 edge!

4. Cut between the numbers and the pinholes all the way around. What is the surprise? Is either piece still a Mobius strip? Explain.

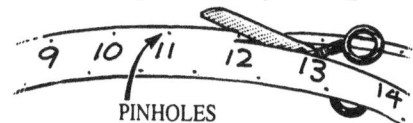

PINHOLES

5. Carefully tear the Mobius strip through the center of all its numbers. What happens? Is the torn piece still a Mobius strip? Explain.

© 1994 by TOPS Learning Systems                    18

---

cards 17-18

## POSTAL BAR CODES ○ Math Lab ( )

1. Draw vertical lines on lined paper to form a grid 10 rows deep and 5 squares wide.

   a. Fill in the top row with 2 tall and 3 short bars like these.

   b. Fill in the remaining 9 rows, always with 2 tall and 3 short bars, but in *different* sequences.

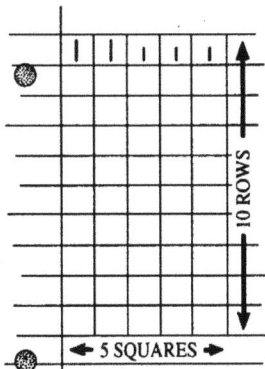

10 ROWS

← 5 SQUARES →

2. Write these numbers above the five columns. Add up the value of the *long* bars on each line (short bars don't count). Write totals to the right.

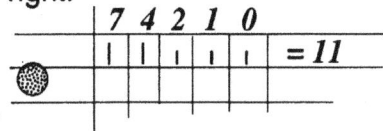

$$7\ 4\ 2\ 1\ 0$$

| | | | | | = 11

3. Get a <u>Bar</u> <u>Code</u>. All codes and their numbers along the bottom match the totals in your table, except one. Which has been assigned a new value?

4. The correction character (last box) is chosen so all the numbers add up to a multiple of 10. Show that this is true.

5. Write correct bar codes for these addresses:

| TOPS Learning Systems | Jo Smith | Your Name |
|---|---|---|
| 10970 S. Mulino Rd | PO Box 345 | Address |
| Canby OR 97013-8799 | LA CA 90027-3618 | City State Zip |

6. Decode bars cut from different envelopes. Label each answer with the circled letter on the envelope.

19

---

## JIGSAW PUZZLES ○ Math Lab ( )

1. Fold the <u>Centimeter</u> <u>Grid</u> in half at the arrows, so all lines match when you hold the paper up to strong light. Stick together with a thin, even layer of glue.

2. Only five of these shapes are unique.

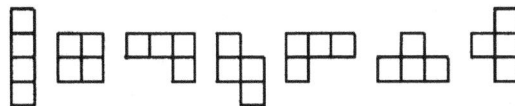

Find 2 shapes that are duplicated…
   a. by rotation.    b. by mirror image.

FOLD

GLUE

3. Cut 12 unique shapes out of your glued squares that have FIVE tiles each. Record the shape of each piece on a new tile sheet.

4. How many different rectangles can you build from these 12 shapes? Record each solution on your tile sheet. (You may turn pieces over.)

20

# Supplementary
# Cutouts
# (four pages)

**ACTIVITY 5**
**GOLDEN RECTANGLES**

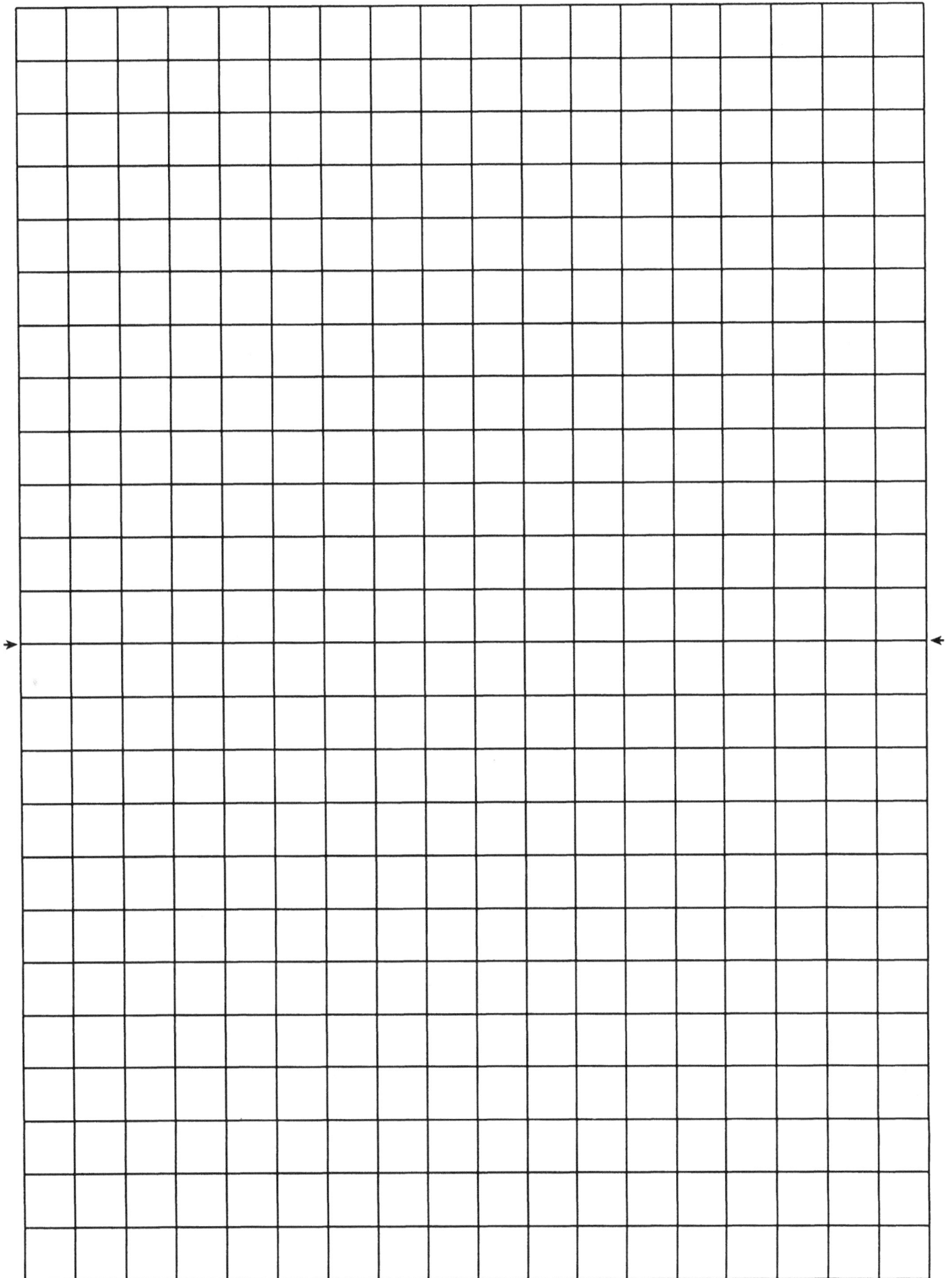

# ACTIVITY 8
# NUMBER SQUARE

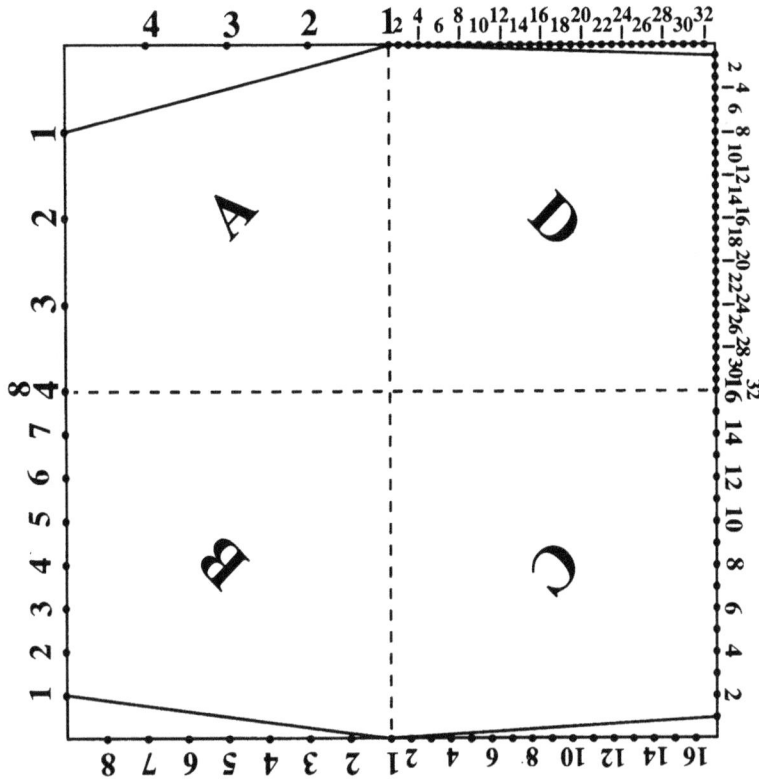

✂ - - - - - - - - - - - - - - - - - - - - - - - - - - - - - - - - - - - - - - - - - - - - - - - - - - - - - - - - - - -

# ACTIVITY 19
# BAR CODE

Any Person

2301 Any St

Any Town US 12345-6789

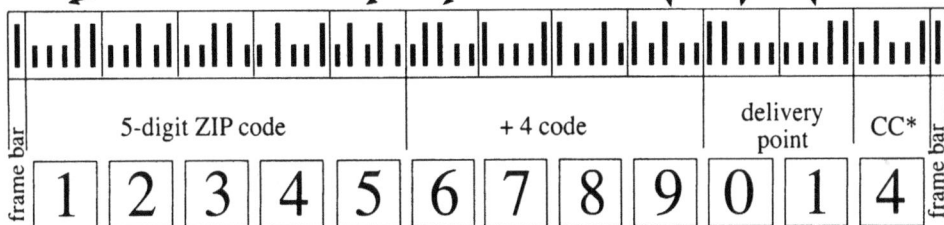

| frame bar | 5-digit ZIP code | | | | | + 4 code | | | | delivery point | | CC* | frame bar |
|---|---|---|---|---|---|---|---|---|---|---|---|---|---|
| | 1 | 2 | 3 | 4 | 5 | 6 | 7 | 8 | 9 | 0 | 1 | 4 | |

* Correction Character: choose this digit so the total of all these numbers is a multiple of 10.

The dots form a circle, numbered around its circumference. Reading outward, the labels include:

1, 38/2, 36, 35, 34/70, 33, 32/68, 31, 66/30, 29, 64/28, 27, 62/26, 25, 60/24, 23, 58/23, 21, 56/20, 19, 54/18, 17, 52/16, 15, 50/14, 13, 48/12, 11, 46/10, 9, 44/8, 7, 42/6, 5, 40/4, 3

--- ✂ - - - - - - - - - - - - - - - - - - - - - - - - - - - - - - - - - - - - - - -

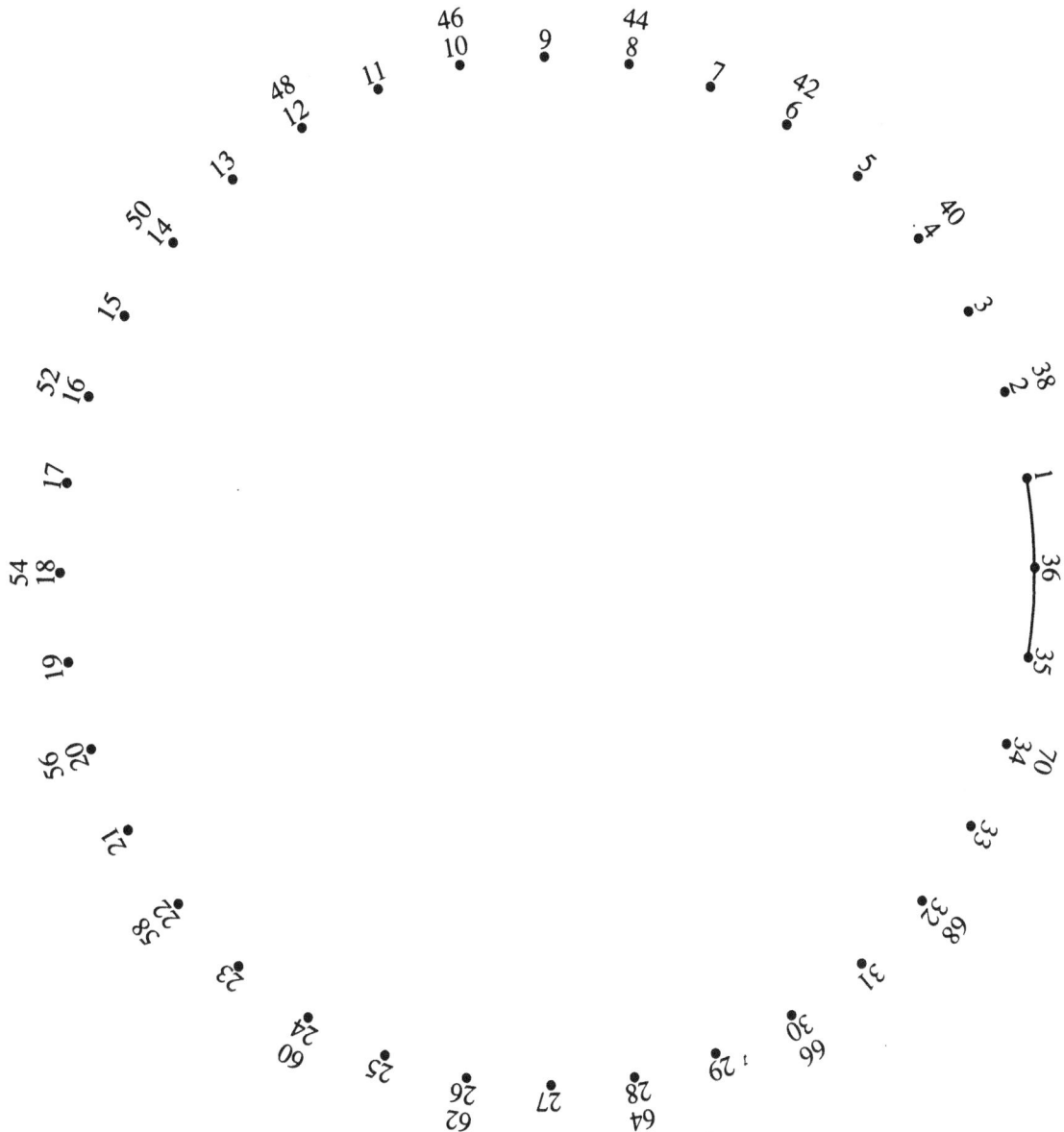

## ACTIVITY 15
## NUMBER SIEVE

The number 1 is neither **prime** nor **composite**.

**Prime** numbers are divisible only by themselves and the number one.

**Composite** numbers are multiples of smaller **primes**.

| 1 | 7 | 13 | 19 | 25 | 31 | 37 | 43 | 49 | 55 | 61 | 67 | 73 | 79 | 85 | 91 | 97 |
|---|---|----|----|----|----|----|----|----|----|----|----|----|----|----|----|----|
| 2 | 8 | 14 | 20 | 26 | 32 | 38 | 44 | 50 | 56 | 62 | 68 | 74 | 80 | 86 | 92 | 98 |
| 3 | 9 | 15 | 21 | 27 | 33 | 39 | 45 | 51 | 57 | 63 | 69 | 75 | 81 | 87 | 93 | 99 |
| 4 | 10 | 16 | 22 | 28 | 34 | 40 | 46 | 52 | 58 | 64 | 70 | 76 | 82 | 88 | 94 | 100 |
| 5 | 11 | 17 | 23 | 29 | 35 | 41 | 47 | 53 | 59 | 65 | 71 | 77 | 83 | 89 | 95 | |
| 6 | 12 | 18 | 24 | 30 | 36 | 42 | 48 | 54 | 60 | 66 | 72 | 78 | 84 | 90 | 96 | |

# Feedback

If you enjoyed teaching TOPS please tell us so. Your praise motivates us to work hard. If you found an error or can suggest ways to improve this module, we need to hear about that too. Your criticism will help us improve our next new edition. Would you like information about our other publications? Ask us to send you our latest catalog free of charge.

For whatever reason, we'd love to hear from you. We include this self-mailer for your convenience.

Sincerely,

**Ron and Peg Marson**
author and illustrator

## Your Message Here:

Module Title _____ Date _____

Name _____ School _____

Address _____

City _____ State _____ Zip _____

———————————————————————————————— FIRST FOLD ————————————————————————————————

———————————————————————————————— SECOND FOLD ————————————————————————————————

RETURN ADDRESS

_____

_____

_____

TOPS Learning Systems
342 S Plumas St
Willows, CA 95988

TAPE HERE